D1029662

SMART MANUFACTURING WITH
ARTIFICIAL
INTELLIGENCE

Jake Krakauer
Editor

Rachel Subrin
Senior Publications Administrator

WITHDRAWN

Published by

Computer and Automated
Systems Association of SME
Publications Development Department
Marketing Services Division
One SME Drive
P.O. Box 930
Dearborn, Michigan 48121

377201

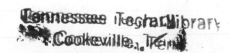

Tennessee Tech Library
Cookeville, Tenn

SMART MANUFACTURING WITH
ARTIFICIAL INTELLIGENCE

Copyright © 1987
Computer and Automated Systems
Association of SME
Dearborn, Michigan 48121

First Edition

First Printing

All rights reserved including those of translation. This book, or parts thereof, may not be reproduced in any form, or by any means including photocopying, recording, or microfilming or by any information storage and retrieval system, without permission in writing of the copyright owners. No liability is assumed by the publisher with respect to the use of information contained herein. While every precaution has been taken in the preparation of this book, the publisher assumes no responsibility for errors or omissions. Publication of any data in this book does not constitute a recommendation of any patent or proprietary right that may be involved or provide an endorsement of products or services discussed in this book.

Library of Congress Catalog Card Number: 87-60008

International Standard Book Number: 0-87263-270-9

Manufactured in the United States of America

SME wishes to acknowledge and express its appreciation to the following contributors for supplying the various articles reprinted within the contents of this book. Appreciation is also extended to the authors of papers presented at SME conferences or programs as well as to the authors who generously allowed publication of their private work.

Aerospace America
1633 Broadway
New York, New York 10019

Computer-Aided Engineering Journal
Institution of Electrical Engineers
P.O. Box 8
Southgate House
Stevenage, Herts
England

I&CS—The Industrial and Process Control Magazine
Chilton Company
Chilton Way
Radnor, Pennsylvania 19089

Journal of Manufacturing Systems
Society of Manufacturing Engineers
One SME Drive
P.O. Box 930
Dearborn, Michigan 48121

Manufacturing Engineering
Society of Manufacturing Engineers
One SME Drive
P.O. Box 930
Dearborn, Michigan 48121

Mechanical Engineering
American Society of Mechanical Engineers
345 E. 47th Street
New York, New York 10017

Robotics & Computer-Integrated Manufacturing
Pergamon Press, Inc.
Maxwell House
Fairview Park
Elmsford, New York 10523

Society of Photo-Optical Instrumentation Engineers
P.O. Box 10
Bellingham, Washington 98227-0010

Test & Measurement World
199 Wells Avenue
Newton, Massachusetts 02159

PREFACE

What is artificial intelligence (AI)? How can artificial intelligence improve productivity in manufacturing, and profitability of manufacturing operations? These questions serve as the impetus for this book. The collection of papers and articles here is designed to give readers an overview of AI applications and detailed descriptions of selected cases where AI techniques are currently in use.

As the Industrial Age wrestles with maturity and approaching seniority, the ability to gain influence in world markets becomes more and more dependent on information. It is not merely possessing information that is important, but rather the ability to organize and access information required to solve specific problems. Human knowledge and experience, if logically filed, manipulated, and recalled by problem solvers, can improve the quality of decisions and shorten the decision-making cycle.

Artificial intelligence is the capturing of experience, knowledge, and information, filing it and arranging it for selective recall. The word *artificial* is a misnomer since the intelligence is not truly artificial, but based on human knowledge and experience. The computer does not actually "think," but instead executes instructions and manipulates data and "coded experience" to present conclusions based on human-supplied criteria. Computer-based AI is knowledge manipulation.

Manufacturing engineers, managers, and technicians make frequent decisions based on existing data and on their knowledge and experience. Manufacturing is a field in which rules of thumb proliferate, where sequences of events lead to logical conclusions, but where finding the optimal rules or sequences can be a complex task. Because AI is a tool with the potential to substantially improve decision-making abilities, it has begun to penetrate functions in the design and manufacturing operations of companies and research organizations.

AI can provide the key to success for the complex computer integrated manufacturing (CIM) systems that are beginning to take shape. Populating factories with automated equipment and systems will not in itself make an operation more efficient. A robot, like a computer, when installed in a location where there is confusion and lack of control, will simply increase the confusion and decrease the control. Knowledge and experience applied to the machine, however, such as smart vision to allow the robot to grasp, orient, and load a part into a machining station, can lead to a more positive result.

This new technology must prove itself in specific areas of design engineering, manufacturing engineering, and business management before it can significantly affect CIM system control. Initial AI activity centered around oil exploration, medical diagnosis, loan and insurance analysis, and maintenance management. Current applications also include engineering design and producibility, manufacturing scheduling and control, robot vision, and quality control.

AI is not without its critics. There are barriers to the acceptance of AI methods, and there are misconceptions and unrealistic expectations. Managers must realize that AI is a technology, not a market or a product. It is a technique for improving the capability of information processing, a tool for decision makers. It is knowledge manipulation.

In addition, expert systems can add value to conventional computing techniques, but off-the-shelf applications packages for turnkey use are difficult to create. Some degree of customization is usually required to make expert systems software useful in specific applications. This process may involve building "applications smarts" on top of an expert model.

Finally, decision makers may tend to defend their "decision turf" by resisting AI methods that threaten to obsolete their abilities. Although they may be justified in their concern, they should also realize that AI tools can improve the quality and speed of their problem-solving tasks and can act as decision assistants or advisors.

Putting aside potential controversy arising from these comments, the objective of this preface is to point out the potential of AI technology applied in manufacturing, and to caution that AI has limitations because of its complexity and youth. Believing that AI can improve decision-making abilities is justified, but believing that computers can "learn" or replace all human problem-solving processes is misguided.

This book is divided into three portions: an introductory section on AI in manufacturing, a series of chapters covering major applications in design and manufacturing, and a final section of case studies. Papers and articles come from a variety of sources, including industry and academia, major conferences and journals, and U.S. and international authors. Articles in the first section provide an overview of AI and typical applications today. The other sections and the case studies offer explanations of specific, functioning projects with references.

Use this book as a reference document, as a tool for raising your understanding of AI in manufacturing, and perhaps as a framework for investigating and planning your own program. You will have to work to understand some of the details, but hopefully you will feel rewarded by your improved knowledge of this promising technology.

I would like to thank all of the companies, organizations, publishers, and authors who gave permission to have their articles reprinted in this volume. Thanks also to the Publications Development Department staff at SME for their assistance in the research and development required in making this book possible.

Jake Krakauer
Silicon Graphics, Inc.

ABOUT
THE EDITOR

Jake Krakauer is Manager of Product Marketing for Silicon Graphics, Inc. in Mountain View, California. Mr. Krakauer markets computer graphics workstations for computer aided design, mechanical computer aided engineering, manufacturing and visual simulation, and animation applications.

Prior to joining Silicon Graphics, Mr. Krakauer was Manager of Technical Markets at Altos Computer Systems, Product Manager at Megatek Corporation, and Consulting Engineer for Metcut Research Associates, Inc. where he developed expert software for manufacturing process planning.

He has a B.S. degree in Mechanical Engineering from the Massachusetts Institute of Technology, an M.S. degree in Mechanical Engineering from the University of Wisconsin, and an M.B.A. from the University of Virginia.

Mr. Krakauer is a member of the Society of Manufacturing Engineers, the Computer and Automated Systems Association of SME, and Robotics International of SME. He is an active member of the CASA/SME Technical Council and participates in planning and implementing activities of the Artificial Intelligence Committee.

He is also a member of the American Association for Artificial Intelligence, the American Society of Mechanical Engineers, the National Computer Graphics Association, and the American Marketing Association.

CASA/SME

The Computer and Automated Systems Association of the Society of Manufacturing Engineers (CASA/SME) was founded in 1975 to provide comprehensive and integrated coverage of the field of computers and automation for the advancement of manufacturing.

As an educational scientific association, CASA/SME has become "home" for engineers, managers, and other professionals involved in computer-based technologies and automated systems. CASA/SME is applications-oriented and addresses all phases of research, design, installation, operation, and maintenance of the total manufacturing enterprise. This book is one example of its wide-ranging activities.

Specific CASA/SME goals are to: (1) provide professionals with a focus for the many aspects of manufacturing which utilize computer systems automation, (2) provide a liaison among industry, government, and education to identify areas of further technology development, and (3) encourage the development of the totally integrated manufacturing facility.

The CASA/SME Technical Council is dedicated to the implementation of computer integrated manufacturing (CIM). It has focused on artificial intelligence as it applies to manufacturing through a committee that develops products to communicate this technology, its scope, applications, and potential. For information on additional artificial intelligence material, contact the Council or Committee through the Technical Activities Division staff at SME.

MANUFACTURING
UPDATE SERIES

Published by the Society of Manufacturing Engineers and its affiliated societies, the Manufacturing Update Series provides significant up-to-date information on a variety of topics relating to the manufacturing industry. This series is intended for engineers working in the field, technical and research libraries, and also as reference material for educational institutions.

The information contained in this volume doesn't stop at merely providing the basic data to solve practical shop problems. It also can provide the fundamental concepts for engineers who are reviewing a subject for the first time to discover the state of the art before undertaking new research or applications. Each volume of this series is a gathering of journal articles, technical papers and reports that have been reprinted with expressed permission from the various authors, publishers, or companies identified within the book. Educators, engineers, and managers working within industry are responsible for the selection of material in this series.

We sincerely hope that the information collected in this publication will be of value to you and your company. If you feel there is a shortage of technical information on a specific manufacturing area, please let us know. Send your thoughts to the Manager, Publications Development, Marketing Division at SME. Your request will be considered for possible publication by SME or its affiliated societies.

TABLE OF CONTENTS

―――――――――――――― CHAPTERS ――――――――――――――

4   ROBOTICS AND VISION

5   FLEXIBLE MANUFACTURING SYSTEMS

6   INSPECTION

CHAPTER 1

AI IN MANUFACTURING

Reprinted from *Manufacturing Engineering*, April 1985

Artificial Intelligence: A New Tool for Manufacturing

Expert systems, vision, robotics—these are just a few of the fields being embraced by artificial intelligence. Yet, its importance and impact, bound to be dramatic, are just beginning to be felt

BY RICHARD K. MILLER

Although the commercialization of artificial intelligence (AI) began only about two years ago, the progress in developing applications for business and industry has been dramatic. Approximately 100 firms offer AI software and services, and about an equal number of large corporations have established in-house AI departments. It is estimated that more than 100 different AI products, including expert systems and natural language software, are either in commercial use or under development with commercialization scheduled in the next year.

A strong future is forecast for AI. Predictions indicate that half of the computers sold in 1993 will contain artificial intelligence components rather than arithmetic components and will be called "logic machines." By 1993, annual revenues for AI software, hardware, and services are projected to reach $8 billion.

Because definitions can impose limitations, the scope of AI has no formal definition. Basically, AI research includes the wide spectrum of fields that represent the leading edge of computer science. Most AI programs are concerned with symbolic reasoning and problem solving. The various fields of AI technology which are of greatest interest to manufacturing include: expert systems, artificial vision, robotics, natural language understanding, and voice recognition.

The languages of AI

The most widely used AI programming languages are LISP and PROLOG. LISP has been the dominant language in the US, while PROLOG is the favorite in Europe and was selected for the Japanese ICOT project.

AI programming languages differ from conventional procedural languages (FORTRAN, BASIC, PASCAL, etc.) because they allow a programmer to define a desired result without being concerned with the detailed instructions of how it is to be computed.

Sometimes called "declarative" languages, they are well suited to special computers, called parallel processing computers (multiple computers working together simultaneously), as programs that can be run in any order without regard to sequence. The languages which deal with operations, objects, words, and ideas are therefore most adaptable to robotic functions.

Computers for AI

The computers used for AI research for the past several years have been primarily the DEC (Digital Equipment Corp.) System-10 and DEC System-20 family of time-shared machines. These are now being superseded by the more economical DEC VAX time-shared computers and the newer personal AI machines. The newer machines tend to have 32-bit words, sorely needed for address space, as most AI programs are huge.

The Massachusetts Institute of Technology (MIT) Artificial Intelligence Laboratory designed a personal machine especially microcoded for LISP. This MIT LISP machine has been licensed to Symbolics, Inc. (Cambridge, MA) and LISP Machine, Inc. (Los Angeles, CA). Personal AI computers are also offered by Xerox Special Information Systems (Pasadena, CA), PERQ Systems Corp. (Pittsburgh, PA), and Tektronix, Inc. (Beaverton, OR). It is expected that the price of good AI personal computers that run LISP will continue to drop as demand escalates.

Expert systems

Expert systems are currently the most emphasized area in the field of artificial intelligence. Dr. Edward Feigenbaum, director of the heuristic programming project at Stanford University and a pioneer in the field, defines an expert system as an intelligent computer program that uses knowledge and inference procedures to solve problems that are difficult enough to require significant human expertise for their solution. The knowledge necessary to perform at such a level, plus the inference procedures used, can be thought of as a model of the expertise of the best practitioners of the field. The knowledge of an expert system consists of facts and heuristics. The "facts" constitute a body of information that is widely shared, publicly available, and generally agreed upon by experts in a field. The "heuristics" are mostly private, little-discussed rules of good judgement (rules of plausible reasoning, rules of good guessing) that characterize expert-level decision making in the field. The performance level of an expert system is primarily a function of the size and quality of the knowledge base that it possesses.

A human "domain expert" usually collaborates to help develop the knowledge base. Once the system has been developed, in addition to solving problems, it can also be used to help instruct others in developing their own expertise.

It is desirable to have a natural language interface to facilitate the use of the system in all three modes. In some sophisticated systems, an explanation module is also included, allowing the user to challenge and examine the

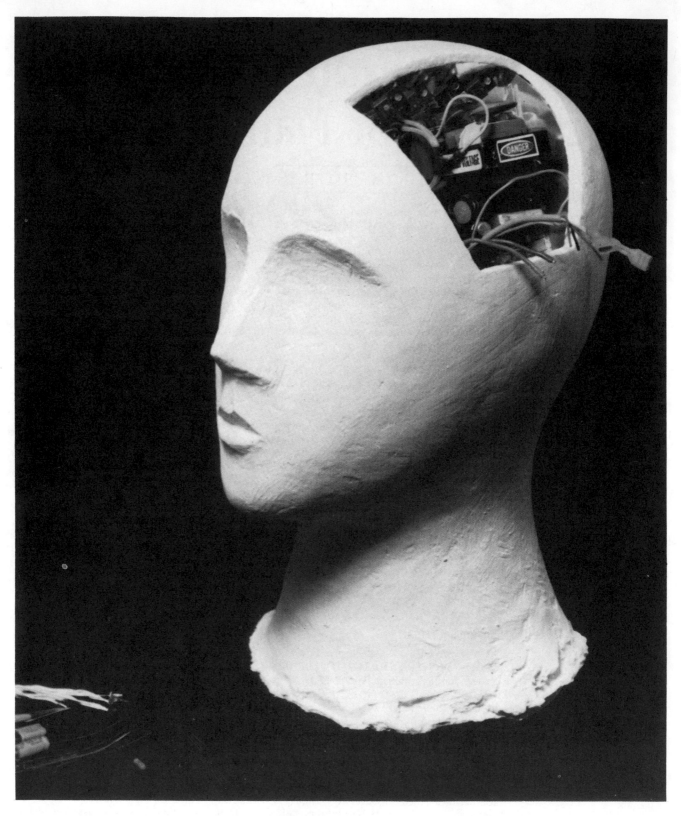

reasoning process underlying the system's answers. *Figure 1* illustrates an idealized expert system. When the domain knowledge is stored as production rules, the knowledge base is often referred to as the "rule base," and the inference engine as the "rule interpreter."

The person who develops an expert system by structuring the knowledge of a domain expert into the AI program is called a knowledge engineer. The knowledge is generally entered into the program in the form of if-then rules. A prototype expert system may have less than 200 rules, while a fully developed program may possess a few thousand rules.

Building expert systems

The time required to develop the early expert systems, often 10-20 work years, was a big drawback to those considering new applications. A large portion of the development

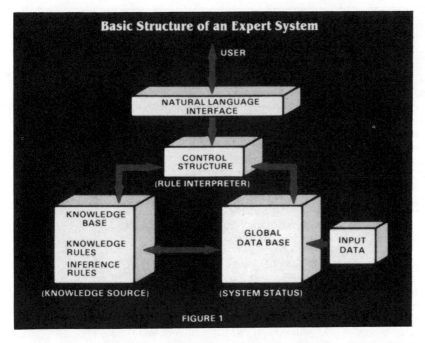

Basic Structure of an Expert System

USER

NATURAL LANGUAGE INTERFACE

CONTROL STRUCTURE

(RULE INTERPRETER)

KNOWLEDGE BASE

KNOWLEDGE RULES

INFERENCE RULES

(KNOWLEDGE SOURCE)

GLOBAL DATA BASE

INPUT DATA

(SYSTEM STATUS)

FIGURE 1

time was consumed in developing knowledge representation schemes and reasoning strategies to structure the inference engine. Since similar reasoning strategies may be applied to problems in numerous domains, AI researchers began to develop domain-independent expert systems, commonly called "tools for building expert systems." By using such tools, the knowledge engineer's task is simplified to eliciting knowledge from an expert and representing this knowledge in a form suitable for the expert system. The laborious task of building the inference engine is eliminated.

In the 1970s and early 1980s, about a dozen such expert system development tools were built by universities and other research groups.

Application of Expert Systems to Design

PRODUCT OR PART SPECIFICATION

INITIAL DESIGN

RULES AND ANALYSES FOR INITIAL DESIGN

EVALUATION CRITERIA

EVALUATION RESULTS

DESIGN

RULES AND ANALYSES FOR REDESIGN

NO — ACCEPTABLE? — YES

FIGURE 2

Among the most popular of these systems were EMYCIN, OPS5, and ROSIE.

In 1983, several AI firms announced the availability of expert system development programs. In early 1984, some of these tools began to become commercially available. The commercial systems are claimed to be easier to use and better supported than the academic systems. The marketing of these programs has great business potential. Several start-up companies expect to build multimillion dollar businesses selling these programs. Competition is already strong among suppliers.

Expert systems for CAD/CAM

Expert systems have been applied to computer-aided design tasks, such as VLSI and electronic circuits, and have been proposed for CAD/CAM applications.

Dr. Margaret Eastwood, director of integrated factory automation at GCA Corp., believes that a certain amount of design checking and analysis will be incorporated into CAD systems using AI methods. A CAD system could alert a designer that a design does not conform to specifications, for instance. She also suggests that an intelligent CAD/CAM system might perform functions such as selecting the inspection points of manufactured products based on an analysis of its design.

Research at General Electric (Schenectady, NY) is exploring an expert system to perform a design primarily by redesigning, using a comprehensive formal method of evaluation. Because many actual problems are in fact redesigns, an initial design is usually available as a starting point. If not, a simple algorithm that interrogates human experts will generate one. The system then iterates to an acceptable design through evaluation and redesign. For each redesign, the system formally evaluates specific characteristics of the current version in accordance with criteria and scales established in consultation with human experts. The program then recommends or actually performs the redesign. The system repeats this process until the design achieves acceptable performance as determined by its evaluation. Knowledge of evaluation criteria and redesign rules is represented in production rules (in simple cases) or in frames and rules (in more complex cases). The approach is outlined in *Figure 2*. Initial applications are for simple mechanical systems such as V-belts and shafts. Future applications extend to such processes as extrusion and redesign rules for geometry and manufacturing.

Expert systems for maintenance

The maintenance of a complex item of equipment involves a diagnostic procedure incorporating many rules as well as judgement decisions by the maintenance mechanic. Experience is a very important factor in determining the ease with which a mechanic can locate a failure problem and implement the appropriate correction. Expert systems are now being utilized to assist maintenance personnel in performing complex repairs by presenting menu-driven instruction guides for

THERE ARE EXPERTS...

The way an expert system is built, generally speaking, is just as you might think: System builders, knowledge engineers, *talk*, often at great length, to human experts—unique individuals who possess a high degree of special knowledge or expertise in a given subject, field, or technology that's needed to do a specific job. The engineers literally draw from the experts what they know and how they use this particular knowledge in problem solving. This is then incorporated in computer programs, making the knowledge and expertise readily accessible.

And, there you have it: a knowledge-based computerized expert system. However, creating the system is one thing; making it a commercial success is quite entirely another matter. And, let us say this, too, lest we leave the impression that creating an expert system is an easy process: Converting the knowledge from human experts to expert systems takes a real specialist, an artist of sorts, and of these there are very few.

DEC's XCON. In 1978, Digital Equipment Corp. (DEC), in a joint development effort with Carnegie-Mellon University, began work on XCON (expert configurer). XCON was conceived as a tool to help DEC configure its VAX computer systems—a complex task as the VAX product line is broad and deep and each order is uniquely tailored.

According to Dennis O'Connor, group manager, intelligent systems, Digital Equipment Corp. (Hudson, MA), the manual process for configuring the VAX systems had begun to grow complicated and pressured, especially as system orders became increasingly complex and order volumes began to grow. "So," says O'Connor, "XCON was developed to help optimize the process and do it more efficiently."

What XCON does, in essence, is to locate components in sensible physical locations and to properly connect all the elements together. One of XCON's tasks, for example, is to decide if an order specifies something that makes sense. XCON asks if the VAX system that's been ordered can be built and if it is supportable. If something isn't right, XCON identifies what it is. It then documents how to correct the problem and issues detailed output for use by manufacturing and field service personnel.

Results. "We've been using it in our plants since 1980," says O'Connor. "It's analyzed roughly 110,000 unique VAX orders and is running at about 98% accuracy." Not a bad record.

The system has more than 4000 rules it checks in configuring VAX orders. Here's a sample: Rule R88:

IF: The current subtask is assigning devices to unibus modules
and there is an unassigned dual-port disk drive
and the type of controller it requires is known
and there are two such controllers, neither of which has any devices assigned to it
and the number of devices these controllers can support is known

THEN: assign the disk drive to each controller
and note that each controller supports one device

What that rule means, and what XCON is asking here, is to put a disk on a particular bus, match up all the components to be sure that they are the right ones, and if they are, assign the bus, the controller, and the disk simultaneously.

What XCON provides DEC with is assurance of shipment of complete and workable VAX systems—verified, built, and installed correctly and efficiently. "XCON gets it right the first time," O'Connor says, "which results in less time spent in system start-up and increased customer satisfaction. Further, there have been fewer changes in orders configured by XCON than in orders configured manually."

XSEL, XSITE. XCON is used to get VAX systems built and installed. Another system, XSEL, is designed to help customers configure their own systems. The idea here is to help a salesperson, using a portable terminal in a customer's office, prepare quotes for DEC systems. XSEL will be, on the whole, more interactive than XCON and will have all the knowledge of XCON and more.

XSITE, on the other hand, deals with environmental problems: ample on-site power supplies, air-conditioning requirements, elevator capacities, entryway opening sizes, cable length needs. "XSITE deals with the physical problems of the location in which the system will be placed," O'Connor says. "XSITE transfers all that information to XCON which then determines if the system can be properly configured given those parameters."

Drawbacks. Expert systems are not problem-free. For one, they tend to be highly memory intensive, which may not be a terribly big problem as the cost of hardware continues to decline. Another certainly near-term problem is that the specialists required to set up such systems are in short supply—estimates suggest perhaps as few as 250 people in the entire US can tackle *serious* expert system challenges. A third and possibly thorny problem is potential resistance to expert systems on the part of end-users and programmers. Human nature being what it is, reluctance to accept, let alone embrace, expert systems can be expected. And last, expert systems can't be accurate all of the time. But then again, neither are *human* experts. **ME**

—Robin P. Bergstrom

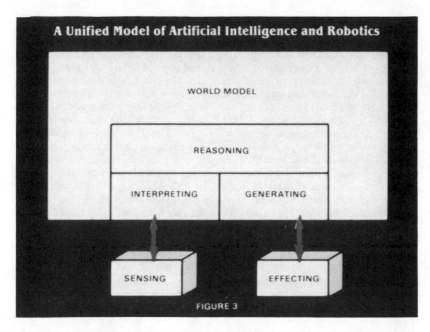

A Unified Model of Artificial Intelligence and Robotics

WORLD MODEL

REASONING

INTERPRETING GENERATING

SENSING EFFECTING

FIGURE 3

the diagnostic task. These expert systems incorporate the knowledge of mechanics who are very experienced in the maintenance and repair of that item of equipment.

The first maintenance expert system to become commercially available was CATS-1 (renamed DELTA). This expert system was introduced by General Electric Co. in August 1983. DELTA (Diesel-Electric Locomotive Troubleshooting Aid) is now being used in railroad "running repair shops" to assist maintenance personnel in isolating and repairing a large variety of diesel-electric locomotive faults. The system was initially named CATS-1 (Computer Aided Troubleshooting Systems—Version 1). To isolate a problem, the troubleshooting system first displays a menu of possible fault areas. When the user selects a particular fault area, the system proceeds with a series of detailed questions. ("Is the fuel filter clogged?" or "Are you able to raise fuel pressure to 40 psi [276 kPa]?") At each step, it follows and explains the expert's reasoning. ("IF engine-set-idle and fuel-pressure-below-normal and fuel-pressure-gage-OK, THEN fuel-system-fault.") At appropriate points during the question-and-answer session, CAD drawings or video disk sequences are displayed on a CRT screen to assist the user in locating various components. Finally, when the troubleshooting system identifies the cause of the malfunction, it generates specific repair instructions (including—if requested—training film sequences displayed on a video monitor).

Expert systems for process control

The rapid advance of computing technology has vastly augmented the capabilities of process control systems. Modern distributed control systems can monitor thousands of process variables and alarms, delivering a constant stream of information to the control room. With such systems, a small crew of operators can regulate the operation of immensely complex industrial processes. Refineries, chemical

processing plants, and power generation plants are just a few of the industries that are benefiting from the rapid growth of this technology. But delivering the data to the control room is only a small part of process control. The incoming data must still be analyzed, understood, and acted upon by human operators. And while their abilities match the demands of day-to-day operation, even the most rigorously trained operators can be overwhelmed by the flood of alarms and upset indications generated by a process interruption or fault.

PIPCON (Process Intelligent CONtrol) is a real-time expert system for process control available from LISP Machine, Inc. PIPCON can monitor up to 20,000 measurements and alarms and can assign priorities to alarms to assist an operator in dealing efficiently with a process interruption or fault.

Other important areas being addressed by expert systems in manufacturing include energy management systems, facilities management systems, and factory and plant design simulation systems.

Artificial vision

Machine vision currently enjoys the position of being one of the most exciting emerging technologies in manufacturing. The field found its birth in several AI research laboratories across the US in the 1970s. While some of the early research results have found their way into the commercial market, research continues today on more advanced problems of artificial vision.

The first successful methodology for vision analysis was the SRI Vision Module, developed at SRI International (Menlo Park, CA) and Stanford University. It was this algorithm which brought computer vision into commercialization in the late 1970s as several companies offered systems based on this research.

The SRI Vision Module was designed to locate, identify, and guide manipulation of industrial parts. The connectivity analysis program breaks a binary image into its connected components. The program builds a description of each blob (a connected component, either an object or a hole) as the image is processed. An array is created to hold information about the blob and its shape. Finally, a number of shape and size feature values characterizing the blob are derived. The connectivity analysis can provide the following geometric information: maximum and minimum values of width and height (X and Y values), area, perimeter, length, holes, centroid position, moments of inertia, orientation, elongation index, compaction index, and a linked list of coordinates on the perimeter.

In another example, two camera vision systems developed at the MIT Artificial Intelligence Laboratory use the same approach as human vision, binocular parallax—the slight disparity between the two eyes' views of a scene—as a depth cue. The greater the parallax, the closer an object. The MIT vision system reduces objects in a pair of binocular images to their outlines and then matches edges by thickness, orientation, and contrast with background. Once this has been done, the depths of the edges can be computed from their parallax and intermediate points interpolated.

Robotics and AI

Artificial intelligence is the area which needs most to be developed and mastered to accelerate robot evolution. An intelligent robot is one capable of receiving communication, understanding its environment by the use of models, formulating plans, executing plans, and monitoring its operation.

Some researchers do not distinguish between AI and robotics; instead a unified model that encompasses both is used. An intelligent robot should be able to think, sense, and effect. Thinking is primarily a brain function. Sensing (seeing and touching) and effecting (moving and manipulating) are primarily body functions. The thinking function executed by a computer is the domain of artificial intelligence. Sensing and effecting are based on physics, mechanical engineering, electrical engineering, and computer science. Planning and execution of tasks entail both brain and body functions and are the concern of both artificial intelligence and robotics. A unified model of AI/robotics, as proposed in an SRI International report for the US Army, is shown in *Figure 3*.

While it is evident to robotic experts that expert systems will provide the technology for future intelligent robots which can cope with unstructured environments, no commercial robots have yet appeared on the market which utilize expert systems. There are indications that leading robot firms may now be giving serious consideration to expert systems for their robots.

At the Robots 8 Conference, Professor Barry Soroka of the University of Southern California discussed the application of expert systems to robotics. Dr. Soroka discussed four types of robotics problems which may be helped by the use of expert systems.

• **Kinematics and design**. For the operation of a robot arm, the kinematic problem must be solved associated with going from a desired Cartesian position and orientation back to the joint angles required to achieve it. An expert system could probably be implemented to replace the human expert in the symbolically tedious work of producing such solutions.

• **Robot selection**. Because of the shortage of experienced application engineers, an expert system could diffuse applications experience among a wider community of users.

• **Workspace layout**. When robots are to be installed in a new workspace, an expert system could provide the optimum workspace requirements.

• **Maintenance**. Expert systems could lead the user through the symptoms of trouble, request test measurements, suggest adjustments and repairs, and monitor the process of fixing the robot.

The development of mobile robots is the subject of research at several AI laboratories. Walking machines, which use one leg for

AI EMPLOYMENT DEMOGRAPHICS

Start-up companies in the emerging artificial intelligence (AI) industry are locating primarily in California and Massachusetts, according to a recent survey by DM Data, Inc. (Scottsdale, AZ), a consulting organization specializing in the AI field.

The survey shows that 90% of all individuals involved in new AI companies are located in only seven states. In order of employment level, they are: (1) California, (2) Massachusetts, (3) New York, (4) Michigan, (5) Florida, (6) New Jersey, and (7) Texas.

A strong correlation between the level of activities in each area and the number of engineering graduates per state was also indicated by the study. AI companies are even more strongly concentrated near the centers for higher learning than are other computer companies. For example, most AI companies in California are located in the area immediately adjacent to Stanford University, and most of the Massachusetts-based AI companies are in the Cambridge area surrounding MIT. Other correlations show a strong relationship between AI activities and the percent of taxes on business versus total state or local taxes. **ME**

—Robin P. Bergstrom

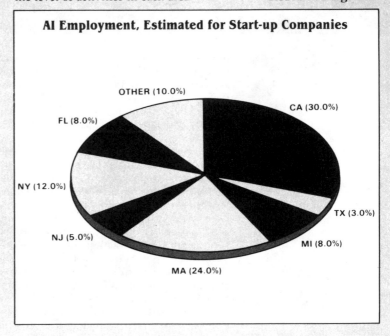

AI Employment, Estimated for Start-up Companies

OTHER (10.0%)
CA (30.0%)
FL (8.0%)
NY (12.0%)
TX (3.0%)
NJ (5.0%)
MI (8.0%)
MA (24.0%)

hopping and two or more legs for walking, are now the subject of research projects at Carnegie-Mellon University and Ohio State University.

Natural language understanding

Much of the intellectual power inherent in computers is untappable by most humans because of a language barrier. Computers operate with machine language—1s and 0s—which is a form of discourse so far removed from normal language that only very small, tortured conversations can take place. This language barrier led to the development of languages that were somewhat more comprehensive to humans but could still be translated into machine language. Many so-called high-level languages were developed and are still being developed. No programming language so far developed, however, is even close to human language. Thus only those who have mastered a programming language have direct access to the power of computers. And master-

Commercial Natural Language Understanding Systems	
COMPANY	**SYSTEM**
ARTIFICIAL INTELLIGENCE CORP. (WALTHAM, MA)	INTELLECT
CARNEGIE GROUP, INC. (PITTSBURGH, PA)	PLUME
COGNITIVE SYSTEMS (NEW HAVEN, CT)	PEARL, MARKETEER, EXPLORER
EXCALIBUR TECHNOLOGIES (ALBUQUERQUE, NM)	SAVVY
FREY ASSOCIATES (AMHERST, NH)	THEMIS
MICRORIM (BELLEVUE, WA)	CLOUT
SYMANTEC (SUNNYALE, CA)	STRAIGHT TALK,
TEXAS INSTRUMENTS (DALLAS, TX)	NLMENU, NATURALLINK

ing a programming language is a difficult, time-consuming task, in contrast, for example, to learning to drive a car.

The ideal form of discourse between humans and computers would be in a language natural to humans—that is, natural language. Much effort has gone into programs that will be able to understand natural language, more effort and resources probably than any other area of AI. If the language barrier can be overcome, computers can start to do many things that today are beyond them. For one thing, they would be able to respond intelligently to human inquiries instead of simply spitting out some prerecorded reply. This would be of obvious advantage in correcting billing errors by stores, in ordering products by telephone, and putting new data into a computer system.

While the problems are difficult, a great deal of progress has been made, at least toward programs that can parse English sentences and pick out nouns, verbs, adverbs, and so forth and answer simple questions about the subject matter. Some programs are capable of understanding simple narratives and can respond appropriately when questioned. One program can process news stories off the Associated Press news wire and pick out key ideas and concepts with about 90% success. Computer programs have great difficulty, however, with unreferenced pronouns such as "it." If one human says to another, "It is raining," the statement is understood. But most computer programs cannot figure out what the "it" is that is doing the raining. Programs also have trouble with analogic statements—"out of sight, out of mind" was once translated by a computer as "blind, insane."

The first commercial natural language understanding system was placed on the market in 1981. Several other products have been commercially introduced since then. The table above presents a list of these systems.

Voice recognition

Voice or speech recognition systems are probably the area of artificial intelligence with the greatest commercial application. It is likely that this form of computer data input will surpass CRT usage at some point in the future.

The voice recognition processes may be classified by their recognition capability. In continuous-speech recognition, word endpoints may be ambiguous (for example, "porous" versus "poor us"), but endpoint determination can be aided by knowledge of the language syntax. In isolated-word recognition, word endpoints are determined by the periods of silence, which typically must last at least 200 milliseconds.

Voice recognition systems which are "speaker independent" are designed to recognize the speech of most speakers. Such systems offer the most flexible potential application. However, within the current state of the art, a higher recognition reliability is achieved with "speaker dependent" systems. These systems are "trained" by the operator by repeating each word in the vocabulary several times. Dialects, accents, or the language itself makes no difference. A Spanish-speaking operator, for example, might choose to train the system in Spanish. Each operator's training utterances are stored on tape while the personal identification code, keyed in before data entry, loads individual training utterances into the system's active memory. There is no limit to the number of operators who can use such a system, even if they all speak different languages.

At this point in time, voice recognition research is conducted somewhat separately from speech understanding research which utilizes written or CRT-input words. Both the problems of recognizing the spoken word and understanding language structure are of sufficient magnitude that they must be attacked separately. As these fields advance, the two technologies will merge. By the end of this century, we may begin to see computers that we can talk to in conversational speech; they will talk to us and do what we ask them to do. For the present time, the primary applications of voice recognition are data entry, NC programming, and robotics.

The future

The commercial application of artificial intelligence is truly in its infancy. Many researchers are concerned that the great public exposure which AI is receiving by the media may lead to unrealistic expectations. This may result in a backlash against the field and deter future funding. While the accomplishments of AI programs are impressive, there are many limitations. Assuming the field avoids a "dark ages" caused by high expectations of current popularity, AI research will continue toward the goals of making computer-based systems more powerful and easier to use. Future generations of AI software and systems will produce machines which can cope with a changing environment and learn from experience. Also in the future will be cognitive computers which can deal with users on a very human level. **ME**

Richard K. Miller, CMfgE, is technical director of SEAI Technical Publications (Madison, GA), a consultant, and author of numerous studies on advanced manufacturing technologies.

Presented at the CASA/SME AUTOFACT 6 Conference, October 1984

Artificial Intelligence—Applications in Manufacturing

by Richard J. Mayer
Don T. Phillips
and
Robert E. Young
Texas A&M University

INTRODUCTION

The techniques and theoretical results from the field of artificial intelligence (AI) offer a new and exciting technology for solving problems in manufacturing today. The technology, however, is still in its infancy. Careful analysis of the characteristics of a problem area must be made to determine if the technology can actually be applied. For maximum benefits to be achieved the AI based applications must be integrated with existing manufacturing systems and practices. Much of the current interest in the area of AI applications to manufacturing has focused on shop floor automation [Gevarter 83, Miller 84]. We believe the greatest payoff for this technology will be in applications to the planning, control, and information management functions. The first two sections of this paper provide some insight into the rationale behind our orientation. The remainder of the paper describes:

a) The kinds of applications which can be made with the AI research results to date.
b) The characteristics which distinguish problems amenable to solution with existing AI technology.
c) Some preliminary results of our ongoing identification of areas in manufacturing planning and control which are candidates for the application of this technology.

The impetus behind this study comes from two sources. The Texas Engineering Experiment Station (TEES) at Texas A&M has initiated the establishment of a Manufacturing Systems Science Center. The objective of this center is to pull together traditionally disjoint research efforts in computer science, industrial, mechanical, and electrical engineering to address the problem of manufacturing productivity from a systems perspective. The research reported in this paper is part of an ongoing effort to define the specific manufacturing problem areas, identify technology voids and to define research goals. This work is also supported by the Air Force Materials Laboratory Manufacturing Science Program to provide planning data for an "AI in Manufacturing Research Center" [Russo 84].

THE PROBLEM SITUATION

The capability of American industries to compete in the marketplace at home and abroad has been eroded by many factors. The high rate of inflation (during the late 70's and early 80's), lack of capital improvement incentives, social trends, and the low priority given to manufacturing research and development have all contributed to this erosion.

The manufacturing industry produces 75% of its product in the traditional batch manufacturing environment, authorizing orders for detailed parts 3 to 4 times per year in lot sizes of 30 to 50 parts. This dynamic start/stop environment is further aggravated by the high level of engineering changes and schedule changes, as well as manufacturing technology changes. To counter the above problems, high work-in-process inventory levels have been introduced to enable manufacturing management to better utilize the machine processing unit. Partial success has been achieved, but optimization of the product throughput and cost leadership have suffered. This is primarily due to the lack of effective manufacturing planning and control and to the fact that existing information systems react slowly to change. Work-in-process spends only 1.5% of its time in a value added activity; the other 98.5% of the time is spent in queues, movement, machine setup, quality inspection, etc. Obviously, such a system can not be considered efficient from either the machine utilization point of view, or from the inventory investment point of view. Furthermore, the impact on quality due to the possibility of part damage (in storage or movement) or part spoilage (due to engineering change) has been significant.

The solution lies in improved manufacturing planning and control systems, and the integration of these systems with the design engineering and business management systems. It has been shown that significant improvement can only be made in the above environment by optimizing the total system, rather than optimizing a single machine or process [Morrison 83]. The solution requires use of the power of the digital computer to assist in the integration of the corporate information base and to improve the accuracy, timeliness and effectiveness of management decision making. In this paper we will address the application of a new technology, based on research in artificial intelligence, to this information integration problem.

THE SOLUTION REQUIREMENTS

Central to the solution of the above problems are the concepts of information integration and end-user computing support [Young,Mayer 84]. The concept of information integration embodies the following three factors:

a) Management of information as a corporate resource.

b) Automated enforcement of the enterprise rules for data existence, integrity, relatability, and access.
c) Sharing of common data and common system modules among all engineering, business and manufacturing systems in the corporation.

A measure of the integration in a manufacturing system is its ability to respond effectively and economically to changes in demand, capital, labor, materials, and product or manufacturing processes. The effect of better integration can be shown to accomplish any or all of the following:

a) Improve an individual function.
b) Increase the amount of information (variables) taken into account by a function.
c) Increase the amount of relevant information produced by a function.
d) Increase the capabilities of the information management system to accurately collect, store and retrieve data.

There are two major issues associated with the concept of effective end-user computing:

a) Ready access to facts and data.
b) Modeling and analysis capabilities.

The first issue concerns providing the end-user with ready access to facts (or data) and allowing simple manipulation of that data. Included in this area are requirements for:

a) Interactive query capabilities.
b) Natural language interfaces.
c) Text editing / graphic display generation.
d) Calculator style arithmetic manipulation.
e) Matrix style arithmetic manipulation such as is provided in a typical spread sheet tool.
f) Report generation and business graphics.
g) Collection of (and access to) knowledge based data.

The second issue concerns providing the capabilities to perform modeling and analysis of the following types:

a) Simulation and optimization models.
b) Planning models.
c) Simple inferencing.
d) Creation of decision scenarios which involve linking together several modeling or analysis tools.
e) Diagnosis models.
f) Prediction / forecasting models.
g) Project management models.
h) Design analysis models.
i) Access to the above capabilities with a simple and easy-to-use interface.

The advantages associated with the provision of these end-user computing facilities include:

a) Reduction in user support applications development time.
b) Providing the user with the capability to utilize common data in unique ways which improve individual performance.
c) Provide mechanisms for capturing the organizations knowledge base.
d) Provision for rapid technology transfer within and between organizations.

Initiatives into either of the above outlined areas are complicated in the manufacturing environment by the inherent requirement for integration with existing systems. Figure 1 displays a conceptual view of how the existing business, engineering, and manufacturing systems would be supported by an information integration mechanism and augmented by the knowledge based end-user computing facilities.

THE ROLE OF AI

The discipline of AI since its inception in the 1950's has been concerned with understanding how humans acquire, organize, and use information. AI researchers have used the computer as the primary test environment for evaluation of proposed theories. Data structures and algorithms which process the information in these data structures have been designed to explore each new theory. These designs are then implemented on digital computers. The effectiveness of the proposed theory is evaluated by execution of the program and a comparison of it's results with the output of human subjects performing similar tasks.

From the point of view of the solution to the manufacturing productivity problem, AI technology can provide two major contributions. The first contribution comes from the insight and understanding which AI theories provide us as to how people do planning, resource allocation, and general problem solving. By evaluating the AI results in these areas in the context of the manufacturing environment, we can develop a better understanding of the type and form of information which must be provided to improve the accuracy, timeliness, and effectiveness of manufacturing planning and control activities.

The second (and probably the more visible) contribution of the AI research to date comes from the potential for direct application of the computer algorithm and information representation schemes directly to manufacturing problems. These applications fall into three major classes:

a) Those applications which attempt to duplicate natural human abilities (vision, language processing, etc.).
b) Those applications which attempt to duplicate learned skills or expertise.

c) Those applications which simply improve the flexibility or effectiveness of traditional information management systems.

The first class of applications include natural language processing systems for:
   a) command interpretation
   b) speech recognition
   c) data base query processing
   d) documentation generation
   e) office message management

All of which directly support the goal of improved end user computing. This class of applications also includes the use of vision and the fusion of vision information with other sensory information to improve:

   a) robot control,
   b) adaptive machine control,
   c) automated inspection systems,

as well as providing the technology base for improved safety devices.

The second class of applications includes those applications which are commonly referred to as "expert systems." These applications are concerned with the automation of tasks that are normally performed by specially trained or talented people. The manufacturing goals for this type of automation would include:

   a) Improving the capabilities of the experts themselves.
   b) Capturing the knowledge base of experts who are retiring.
   c) The capability to quickly transfer critical knowledge and skills throughout the organization.

Expert systems (ES) applications should be differentiated from pure AI research because the primary goal of such applications is not the understanding of the basic mechanisms used by the human expert to arrive at a result. The primary goal of the ES applications is to consistently duplicate the results of a human expert. Although the developers of such systems may start from a particular AI theory, the eventual expert system structure may be based totally on traditional decision table, operations research and database technology. As a consequence, the construction of expert systems is often considered an engineering activity and not a part of the AI science mainstream.

The third class of applications may, in fact, offer the greatest potential for savings and increased productivity. However, applications of the results of the AI research in the manner proposed by this third class of applications is not likely to generate a lot of publicity or fanfare. A typical application of AI techniques which would fall into this class is best explained by considering the following example.

One of the most promising approaches to solving the information integration problem identified above embodies the concept of establishing a company-wide common definition of shared information and actually using this definition to control the access, update, transformation, and management of the corporate common data. This concept, or approach, is embodied in the ANSI/SPARC [ANSI 75] notion of a three schema information system architecture, as well as in the ISO conceptual scheme based system architecture [ISO 81].

The Air Force ICAM Program, has developed a running prototype of such a system (known as the Integrated Information Support System, or "IISS"). The IISS stores the enterprise common data definition as well as the definition of the data location, structure and administration rules in what is called the "common data model." Application development and execution in the IISS is automatically assisted and controlled by the information contained in this model [Hurlbut 83]. Another prototype system is the NBS Automated Manufacturing Research facility (AMRF). The AMRF hierarchical control system is being designed around a similar "conceptual schema" approach [McLean 83]. To date these systems have been successful in using this corporate definition to control the access and transmission of common data. However, the automated management of the update process, with the associated data currency, accuracy, and security problems, has remained an elusive goal. The primary problem is the complexity of the rules which govern the update process, and the variety of situations in which the rules must be enforced. The magnitude of the problem is obvious from traditional applications in which 80 percent of the developed code is required not by the primary function which the application was meant to perform, but by the need to validate and enforce the appropriate data management rules.

Research results by Kolwalski, et. al., suggest that one possible solution to this problem would be to utilize a "logic based" representation of the common data [Kowalski 78]. Variants of algorithms which have been designed for theorem proving in AI planning and problem solving systems could then be employed to automatically provide for the enforcement of the rules and for the appropriate modification of those rules as the situation demanded. This kind of AI technology application to the manufacturing information integration problem would not outwardly appear to embody the information system with any "human-like" characteristics. On the other hand, effectively solving this update problem is one of the primary requirements for significant productivity improvements in manufacturing. An additional benefit, of a potential significant savings in the time and cost associated with new application developments, would be attractive from a cost savings point of view and from the flexibility it would provide to a company's information management system.

Of the three major AI application classes discussed above, the expert systems or "ES" is the one we will focus on in the remainder of this paper.

RATIONALE FOR SELECTING EXPERT SYSTEM APPLICATION AREAS

Several factors must be considered in the determination of whether a particular manufacturing problem area is amenable to automation using ES technology. These factors include:

- Problem characteristics
- Organizational goals
- Usage environment
- Development constraints

In the preceding section we identified the primary benefits of expert systems as:

a) Improving the capabilities of the experts themselves.
b) Capturing the knowledge base of experts who are retiring.
c) Acceleration of the transfer of critical knowledge.

Many of these benefits have been partially achieved through traditional Group Technology (GT), Decision Support (DSS), or Management Information Systems (MIS) technology.
In fact, depending on the situation, these traditional technologies may offer better and more effective technical solutions, as well asmore cost effective solutions. Figure 2 displays the problem areas which have been addressed by expert systems to date. A quick review of this list immediately points out the fact that GT and DSS systems have been built to address the interpretation, prediction, design, planning, instruction and control problem areas. So it is evident that merely listing a general problem category name is not sufficient for characterizing manufacturing problems which are amenable to ES technology.

Figure 3a and 3b do provide some insight into the characterization problem. These figures display the characteristics, features and capabilities of ES in contrast with those of DSS, GT, and MIS systems. The most striking difference between ES applications and the other three technologies is the ability of the ES to address problems which do not have a well-defined model or algorithmic solution. In the DSS literature this type of problem is often referred to as an "unstructured" decision problem [Keen & Morton 78]. A second major difference is in the ability of an ES to not only use a rule base for reasoning within the problem domain, but also the capability of the ES to reason about its own inferencing process and provide rationale and justification for the conclusions it reaches. The third difference is more subtle and actually refers to the paradigm under which an ES is constructed. In the construction of an ES, the developers actually attempt to capture and represent the decision-maker's knowledge. Building an ES application is primarily a task of construction of a knowledge base, rather than the writing of a program. Thus constructing an ES intuitively embodies the notion of imitating the "way" the human expert performs his tasks. This last factor may explain the growing interest and enthusiasm surrounding ES technology. In fact, even within the OR/DSS discipline it is recognized that those techniques (such as simulation) which are less mathematically oriented (based rather on a user's logical perception and reasoning) are by far the more successful and widely used in practice.

SUITABLE PROBLEM CHARACTERISTICS

One of the most challenging tasks associated with the implementation of ES technology in manufacturing is the appropriate selection of the problem. Because of the venacular which is used to describe the capability of ES technology (intelligence, reasoning, knowledge base, etc.) it is easy to presume that the systems have general human-like capabilities. It should be remembered that the criteria which have evolved to determine the suitablity of a particular problem for ES solution are the result of as many failures as successes with this new technology. The following is a summary of the most important of those characteristics, with a brief explanation of each one.

a) Existence of a human expert is crucial.

A common misunderstanding is that ES technology can generate expertise where none exists today. Current ES technology limitations restrict the problem domains to those which require human expertise. In fact, to a large extent the problem domain is limited to those which can be solved with the expertise of a single human expert. In this light consider the difference between a process planning task and a factory level production requirements scheduling task. It is reasonable to expect a single seasoned planner to accurately assess the relevant characteristics of a part and, based upon his experience and knowledge, select a sequence of operations and an initial raw material form which will result in a correct plan for producing that part. An ES based process planner could perform this task and, as well, improve the overall consistency of the generated plans and allow for easier accommodation of new process or material technology. On the other hand, for any reasonable product configuration and factory capability, the determination of the product requirements and then scheduling those requirements so as to minimize resource contention, maximize through put, and minimize inventory investment has been shown by experience to be a task which humans perform poorly. Thus, while it would be highly desirable to have an "expert" to perform this task, it is unlikely that we can train one, much less, build one.

b) The knowledge base should be reasonably bounded and preferably domain specific.

Problem tasks which require "good shop floor sense" or "common sense", or which require researching material from technical libraries and "handbooks", are generally not appropriate for ES application. Thus, you might say that tractable ES problems are those which are "not too easy", and yet "not too hard". So while the factory production requirements scheduling task does not appear to be a good candidate, the determination of the first article schedule, and the generation of the acceleration curves are tractable problem tasks for an ES. This is because the construction of these schedules is based primarily on knowledge of contract delivery requirements, gross product characteristics, and rules for relating those characteristics with past factory performance. Similarly, resource management within a small segment of the factory (such as a GT or functional cell) which utilizes knowledge of the machine, tool, product, and worker characteristics to maximize performance, during crisis situations, to meet the factory production schedule could also be considered a reasonable candidate task.

c) The performance of an expert on the problem is measurably better than that of a beginner or apprentice.

The successful solution of the problem by the expert should be significantly better than a solution generated by an amateur. The success should be due to acquired mental skill or knowledge gained through experience. Thus, expertise which is based on established interpersonal relations, or just familiarity with the problem situation, is not generally able to be duplicated. This criteria can be extended to also include the fact that there should be evidence that the master's skill is regularly transferred to novices.

d) The problem decisions require consideration of a variety of alternatives and may involve uncertainty.

The decision alternatives should be numerous enough that even the expert generally is not immediately aware of all of them. In addition, the factors which determine the appropriateness of an alternative may involve uncertainty. The expert should be able to (while he may not on a practical basis) assign degrees of support between the evidence and possible conclusions.

e) The number of entities, their relevant attributes, and the useful relationships which define the alternatives or decision factors are bounded.

This characteristic basically states that the things which the ES must manipulate (objects, events, characteristics, etc.) in order to represent and use knowledge are limited in number. It also implies that the existing expert (with some coaching) can identify which characteristics of which entities, and what relationships between those entities are relevant to the problem. In terms of hard numbers the bounds translate into; 50-400 entities with 5-6 attributes and 3-5 relationships maximum. Another way to characterize this aspect of the problem is to try to determine if the task is decomposable into individual independent tasks which require generally less than 5-10 minutes of actual thought by the expert.

From a practical point of view there are several other characteristics which should be taken into account when sizing up a manufacturing problem for ES application. In terms of the organizational goals mentioned above, the problem should be studied sufficiently to be able to ascertain that:

a) Conventional MIS, GT, or DSS technology is not applicable. (Why accept more risk and potentially greater cost?)
b) The desired solution is one which mimics the "best" expert. (Do you really want to make the same errors only at faster and with more far reaching effect?)
c) The organization can live without that expert for the duration of the development effort. (The apprentice won't do!)
d) The solution to the problem represents significant value to the organization.
e) The problem has a recognized easy version that has value to the organization and which can be used as a prototype for evaluation.
f) The organization is amenable to the use of outside expert consultants in the initial portion of the project.

The considerations of the development constraints and usage environment actually go hand in hand. While it is certainly true that the eventual system will (or should) be very natural and easy to use, the process of building ES applications is poorly understood, at best. The systems tend to evolve over time as the experts "debug" and expand the knowledge base. Thus, if the users are accustomed to a more formal development process with well defined requirements up front, they may not have the patience to see the system through its development cycle.

APPLICATION AREAS UNDER INVESTIGATION

Using the above-outlined criteria, we are currently performing an assessment of the following functional areas of manufacturing to identify potential areas for ES applications.
a) Plan for Manufacturing
b) Make and Administer Schedules and Budgets
c) Plan Production
d) Engineering Change Management
e) Obtain Manufacturing Materials
f) Control Production Orders
g) Control Production Items and Tools.

The definitions for these functional areas were derived from the ICAM Architecture of Manufacturing [MFG0 81].This document, along with the requirements documents for several of the major manufacturing systems which were developed under the ICAM Program, provides a detailed description of the functions which have been validated by 20 to 30 manufacturing companies. The descriptions and detailed models of the manufacturing decisions associated with these functional areas are being extracted from the needs analysis and requirments documents produced under the ICAM Decision Support (IDSS) [IDSS 80, 82] and Group Technology Support (GTSS) [GTSS, 81] projects. Figure 4 displays an initial list of those decisions or tasks which meet the criteria described above. This list is by no means complete, however, it does provide examples of the type of problems in manufacturing which can be addressed with ES technology.

In addition to the functional analysis we are also performing an evaluation of existing manufacturing support software systems. The purpose of this analysis is to determine where existing system functionality or user interface could be augmented with the use of AI technology. The following system types are being studied:

a) MRP II type production planning and control systems with "best fit" resource loading capability.
b) Shop floor control and data collection systems.
c) Group technology application development support systems.
d) Manufacturing decision support development systems.
e) Manufacturing cost for design evaluation support systems.
f) Information integration systems designed to provide data integration across networks of heterogenous computers and data base managers.
g) Manufacturing software development and prototyping systems. to provide data integration across a network of manufacturing computers and data bases.

SUMMARY

Expert systems certainly provide a technology base which will prove invaluable to future productivity improvements in manufacturing. We have presented an argument for the position that the most valuable of these applications will be those which address manufacturing planning control and information integration functions. We have also attempted to outline the characteristics of the types of problems for which ES technology is most appropriate.

ABOUT THE AUTHORS

Mr. Richard Mayer is a research associate at Texas A&M University currently pursuing his doctorate in Industrial Engineering. He is doing research in the area of knowledge acquisition tools and automatic model generation from manufacturing system specifications. Dr. Robert Young is an associate professor of industrial engineering. He currently manages the Industrial Automation Laboratory. His research activities and those of his students range from real time control systems for FMS and robotics applications, to parallel processing computer architectures for simulation machines, and expert systems for manufacturing planning and control. Dr. Don Phillips is a full professor in industrial engineering. He currently manages all of the funded research activities in the industrial engineering department. He is also heading up the establishment of the manufacturing system science center. His primary research areas are focused on simulation applications, simulation language development, and closed loop manufacturing planning and control applications.

The authors would like to express their appreciation to Mr. Murali Krishnamurthi for his critique and suggestions.

BIBLIOGRAPHY

ANSI/X/SPARC, Study Group on Data Base Management Systems: Interim Report 75-02-08, 1975.

Cambel, J. A. Implementations of Prolog. Halsted Press, 1984.

Charniak, E., Riesbeck, C. K., McDermott, D. V. Artificial Intelligence Programming. Lawrence Erlbaum Associates, Publishers, 1980.

Fiegenbaum, E. A., Cohen, P. R., Barr, A. "The Handbook of Artificial Intelligence". HeurisTech Press, 1982.

GTSS, "ICAM Group Technology Characterization Code (GTCC)". Ninth Quarterly Report. January, 1981.

Hayes-Roth, F., Waterman, D. A., Lenat, D. B. (eds.) "Building Expert Systems". Addison Wesley, 1983.

Hurlbut, M. R. Integrated Information Support System (IISS). Proceedings of Seventh Annual ICAM Industry Days. June, 1983.

IDSS-"IDSS Baseline Functional Requirements". Interim Report AFWAL/MLTC 8202. February, 1980.

ISO TC97/SC5/NG3 Concepts and Terminology for the Conceptual Schema. Interim Report. February, 1981.

Kowalski, R. Logic for Data Description. Logic and Data Bases, edited by Gallaire and Minker. Plenum Press, 1978.

McLean, C. Mitchel M., Barkmeyer, E. A Computer Architecture for Small-Batch Manufacturing. IEEE Spectrum. May, 1983.

MFG0, 'CAM Architecture Part II'. Final Report, AFWAL-TR-81-4023 IDSS-"IDSS Baseline Functional Requirements". Interim Report AFWAL/MLTC 8202. February, 1980.

Miller, R. K. Artificial Intelligence: A New Tool for Industry and Business. Technology Insights, Inc. Fort Lee, New Jersey, 1984.

Nau, D. S. Expert Computer Systems, and Their Applicability to Automated Manufacturing. National Bureau of Standards Report NBS/R81-2466. 1982.

Nillson, N. J. Principles of Artificial Intelligence. TIOGA, 1980.

Russo, V. J. Plan for an AI in Manufacturing Research Center. Sigart AI in Dayton Conference. April, 1984.

Winston, P. H. Artificial Intelligence. Addison Wesley, 1984.

Young, R. E., Mayer, R. J. The Information Dilemma. Industrial Engineering. September, 1984.

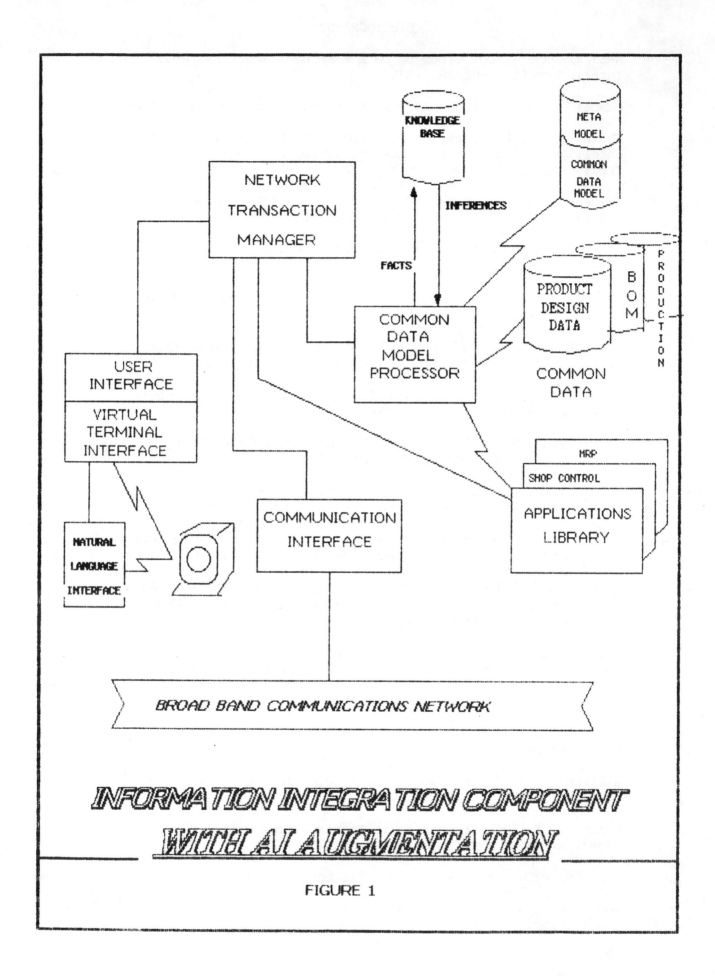

KNOWLEDGE
BASE

META
MODEL

COMMON
DATA
MODEL

NETWORK
TRANSACTION
MANAGER

INFERENCES

FACTS

PRODUCT
DESIGN
DATA

B
O
M

P
R
O
D
U
C
T
I
O
N

USER
INTERFACE

COMMON
DATA
MODEL
PROCESSOR

COMMON
DATA

VIRTUAL
TERMINAL
INTERFACE

MRP

SHOP CONTROL

NATURAL
LANGUAGE
INTERFACE

COMMUNICATION
INTERFACE

APPLICATIONS
LIBRARY

BROAD BAND COMMUNICATIONS NETWORK

INFORMATION INTEGRATION COMPONENT

WITH AI AUGMENTATION

FIGURE 1

TYPICAL EXPERT SYSTEM APPLICATION AREAS

DESIGN

CONFIGURATION

DIAGNOSIS & PRESCRIPTION

EXPERT SYSTEM DEVELOPMENT

INTERPRETATION AND PREDICTION

PLANNING/SCHEDULING/RESOURCE ALLOCATION

FIGURE 2

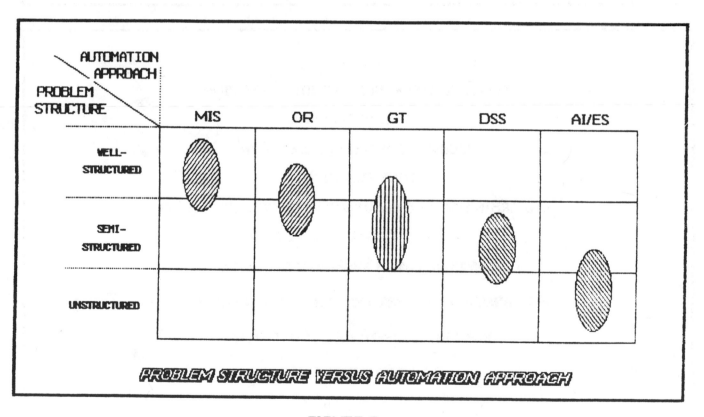

PROBLEM STRUCTURE VERSUS AUTOMATION APPROACH

FIGURE 3a

	MIS	OR	GT	DSS	AI/ES
DEVELOPMENT TIME/COST	MEDIUM	LOW	MEDIUM	LOW	HIGH
APPLICATION EXAMPLE	DATA COLLECTION REPORTING PLANNING ACCOUNTING	FACTORY LAYOUT OPTIMIZATION SIMULATION	PROCESS PLANNING PART FAMILY ID SCHEDULING	DESIGN ANALYSIS RESOURCE ALLOC PROJECT MGT	CONFIGUATION DIAGNOSIS PLANNING DESIGN
EASE OF MODIFICATION	HARD	MODERATE	MODERATE	EASY	EASY

CHARACTERISTICS OF AUTOMATION APPROACHES

FIGURE 3b

QUALITY ASSURANCE METHOD SELECTION

PROCESS PLANNING

TENATIVE MAKE/BUY DECISION

TOOL PLANNING

DETERMINATION OF MANUFACTURING CONFIGURATION

REWORK DISPOSTION REVIEW

DEVELOP BATCH ASSEMBLY SEQUENCE

PREPARATION OF FIRST ARTICLE DELIVERY SCHEDULE

PLAN LINE ASSEMBLY INSTALLATIONS

PLAN FACILITIES LAYOUT

EXAMPLE MANUFACTURING PROBLEMS

SUITABLE FOR ES APPLICATION

FIGURE 4

Reprinted with permission from *Robotics & Computer-Integrated Manufacturing*,
Vol. 1, No. 3/4, Copyright 1984, Pergamon Press, Ltd.

THE DIRECTION OF FUTURE DEVELOPMENT AND THE ROLE OF KNOWLEDGE IN MANUFACTURING TECHNOLOGY

KAZUAKI IWATA and TOSHIMICHI MORIWAKI

Kobe University, Rokko, Nada, Kobe, Japan

The directions of technological innovation and possible future development of the manufacturing technologies are examined, taking into consideration the environments surrounding the manufacturing technologies. The subjects for development in future manufacturing technologies are discussed, and in particular, the roles of artificial intelligence and knowledge engineering are examined in detail. Applications of knowledge engineering to actual problems in manufacturing are introduced, including application of knowledge engineering to the representation and understanding of the specifications in the design stage, knowledge-based process planning, and high intelligence FMS. A new concept called EA (enterprise automation) is also proposed, which deals with the integrated software system of the entire enterprise.

1. INTRODUCTION

In the last decade, development in the manufacturing technologies has been quite remarkable in many of the industrialized countries. As typical examples of the recent innovative technologies, one can mention the CAD/CAM system, FMS, industrial robot, sensing system, local area network (LAN), and so on. The trends of these developments show that the manufacturing technologies are beginning to play important roles when one considers corporate strategies.

Manufacturing technologies are expected to develop further in the remainder of this century, and it is important to know how they have changed from the past to the present in order to predict future trends. They have been supported by various element technologies which have progressed in different ways. Some of them have progressed in a continuous manner, such as the machine tool technologies for instance, while others have progressed in a discrete manner, e.g., new materials, etc.

Needless to say, much effort is devoted to basic and applied research before the individual element technologies emerge and are utilized in practice. The changes in the research and the applications of these technologies are illustrated schematically in Fig. 1. Here, such trends are called the waves of development.

Let us take the case of CAD as an example. Research on CAD began in the late 1950s, and at MIT a research project was started in 1959. This project gave much expectation to many researchers, and the research developed rapidly. As results were obtained, the technical problems and limitations of the possibilities were recognized and the research efforts diminished. However, practical applications of the results of the research were tried in industries, but the problems arose again, limiting the practical use of the technology.

There is a time difference between the two peaks of the waves of the basic research and practical application. Although the time span differs depending on the kind of technology, a typical span is 15–25 years. Some recent technologies have a time span of even 10 years. As the first wave of research decays, it is normal that a second wave of research arises to overcome the problems of the first one. The research results at this stage are reflected in the second wave of practical application. The period designated as the D-point in Fig. 1 is quite important from the practical point of view.

If one looks for some practical technologies in the early 21st century, he should study the basic research occurring at present, taking into account the aforementioned time span. This means that one should concentrate on present-day basic research in

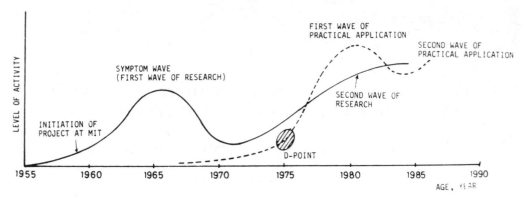

Fig. 1. Waves of basic research and practical applications in technology innovations. (For the case of CAD.)

order to generate the practical technology which will blossom early in the 21st century.

The above example shows clearly that the vital technologies of manufacturing must be cultivated based on a long-range view. In this paper, the environments and constraint factors of the manufacturing technologies of the future are examined first, and then the direction of innovation and the subjects for development are discussed. The roles of artificial intelligence and knowledge engineering are analysed; these are expected to give a large impact to the development of manufacturing technologies in the future. Applications of these technologies are examined regarding the representation of required specifications in designing, process planning in manufacturing software, and intelligent FMS as the manufacturing hardware. The concept of an integrated system of enterprise, called EA (enterprise automation), is proposed as one of the highly intelligent manufacturing software systems of the future.

2. THE ENVIRONMENTS SURROUNDING MANUFACTURING TECHNOLOGIES AND THEIR CHANGES

It is necessary to understand the fundamental environments surrounding the manufacturing technologies in order to examine the manufacturing technologies of the future. The fundamental environments are, as shown in Fig. 2, social environment, natural environment, international relation environment, labor environment, economic environment, science and technology environment, etc.

Changes in the social environment in recent years have been characterized by the wide variation of needs; the transition in the expression of personality, in the recognition of the value of tradition and culture, and in the relative relations between the public and personnel; higher education; etc. As regards the natural environment, it is necessary to consider the

limitations of natural resources, such as energy, harmony with nature, the recycling system of nature, etc. As far as the international relation is concerned, such factors as the harmony between developed and developing countries, among developed countries, and among developing countries; international cooperation; changes in globalism; etc., are to be considered.

The specific features of the changes in the labor environment are the world-wide increase of the population, the geographically biased concentration of the increased population, changes in the composition of workers' ages, shortening of the working hour, and changes in the quality of work. The latter is characterized by the shift of major labor from primary industry (agriculture and mining sector of industry) to secondary industry (manufacturing and construction sector), and further to tertiary industry (service sector). A new sector of industry, which is to be called the quaternary industry, is expected to be generated. Among these trends, the population problem is expected to be the most influential factor.

Changes concerning the economic environment are the low rate of economic growth, unemployment, inflation, rivalry among different kinds of industries, rivalry among the same kinds of industries, generation of new industries, free market and protective market, changes in economic values, etc. Remarkable changes are arising concerning the science and technology environment. These are the developments in computers and information processing technologies; the developments in hardware technologies, such as materials and machines; the developments of ultimate technologies related to space and deep oceans; the generation of new energy; the development of genetic engineering, etc.

It is to be noted here that the items mentioned above as the specific features in the various environments to be considered are limited to the representative items only; many other items should

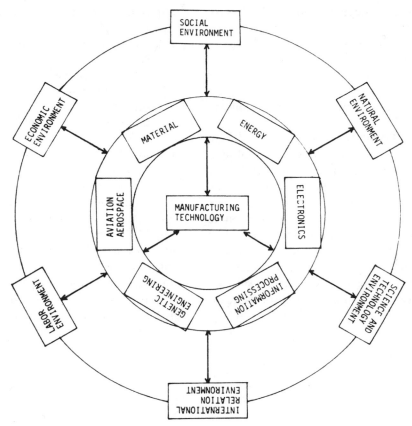

Fig. 2. Fundamental environments surrounding manufacturing technology.

be considered for detailed analyses. It should also be noted that these items are not independent but are mutually related.

In examining the trends in the technology of design and manufacture of products, one must consider especially the changes in the sense of value of customers among the changes in the various environments mentioned above. How will the sense of value be changed in the near future?

The sense of value in the near future has been examined by many people and organizations.[1.5] Briefly, the conclusions are that the value will vary and that harmony with nature is important. The backgrounds of these changes are the desire for a various and high level of life, the maturity of consumers, and the appearance of "prosumers" (producer and consumer).

The products which satisfy the various senses of value of humans are various. Development of technologies for rational design and manufacture is required for various products which satisfy the high needs of individuals, considering the constraint factors of the various environments mentioned above. In other words, development of the manufacturing technology is desired to produce products, adapted to the needs of individuals, which are easy to obtain, use, and maintain by the most desirable means for humans.

3. THE DIRECTION OF INNOVATION AND THE SUBJECTS FOR DEVELOPMENT

The major objectives of developments of technological innovations in the field of manufacturing engineering are summarized as follows, taking into consideration the various environments surrounding the manufacturing technologies explained in the previous section:
- cultivation of new innovative technologies;
- increase of added value;
- increase of productivity;
- saving of resources and energies;
- increase of flexibility;
- preservation of environments;
- improvement of safety; and
- adaptation to social and labor environments.

The trends in the manufacturing technologies consist of the following major items:
- development of new materials (material innovation);

- development of new devices and facilities (product innovation);
- development of new process technologies (process innovation);
- development of new systems (system innovation); and
- development of new software (software innovation).

Studies were carried out by experts in the relevant fields in order to extract and analyse the subjects related to technological innovation. The major fields examined were related to the process, the system, and the software innovations. One hundred and two, 52, and 47 subjects were extracted in relation to process innovation, software innovation, and system innovation, respectively, giving a total of 201 subjects extracted. Each subject was evaluated in terms of its importance, and the following seven subjects were selected as the most important subjects at the time for Japan:

(1) realization of a highly intelligent manufacturing system;
(2) highly intelligent robot (with self diagnosis and recovery functions);
(3) ultra-high precision machining;
(4) nano-meter technology production;
(5) utilization of a high-power stabilized laser;
(6) manufacture of parts with a free form surface; and
(7) intelligent micro-sensor with high sensitivity and reliability.

Among these subjects, important directions of technological innovations are included, which are automation, unattended operation, intelligence, high reliability, systematization, synthesis, high accuracy, etc. In particular, intelligence is expected to give a large impact to technological innovation in the future. In relation to intelligence in manufacturing, the roles of artificial intelligence (AI) and knowledge engineering are examined in Section 4.

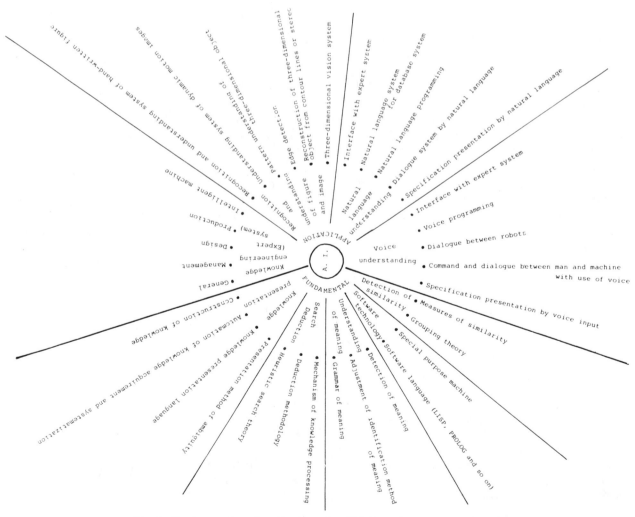

Fig. 3. Fundamentals and applications of artificial intelligence in manufacturing.

4. MANUFACTURING TECHNOLOGIES VERSUS AI AND KNOWLEDGE ENGINEERING

When information concerning manufacturing is processed and designs and operations of the hardware systems are carried out, the knowledge and knowhow of humans (manufacturing engineers and skilled operators), which have been accumulated for many years, are utilized. In this case, knowledge comprises both general and technical knowledge. It also includes the group of knowledge which can be rearranged and utilized according to algorithmic procedures and that which cannot be formulated as algorithms. In decision-making, therefore, the treatment of knowledge which can hardly be formulated as algorithms becomes quite important.

Artificial intelligence deals with such knowledge, and its basis and expected applications in the field of manufacturing engineering contain the items shown in Fig. 3.[3] The expert system which is based on the knowledge, or the new engineering based on the artificial intelligence, is called knowledge engineering. Knowledge engineering is expected to have a wide range of applications; it is also closely related to various fields of manufacturing engineering. Possible applications of knowledge engineering to manufacturing engineering are listed in Table 1. It is understood that knowledge engineering covers a wide range of fields of applications in manufacturing.

In order to apply knowledge engineering to manufacturing software and to construct a suitable knowledge base system, methodologies which suit the manufacturing software system must be developed. The expert system is one of such systems already in existence and is considered here. The basic technologies of the expert system are:

- knowledge representation;
- utilization of knowledge or reasoning;
- knowledge acquisition; and
- man–machine interface.

Knowledge representation concerns the structure and interpretation procedure of data which are used properly within the program to perform intellectual tasks. Research is required to establish the most suitable model for manufacturing systems. Knowledge is used to deduce new facts from known knowledge. It is important to clarify the reasoning process of expert engineers and operators and to establish the most desirable reasoning mechanisms for the manu-

Table 1. Applications and subjects of knowledge engineering in manufacturing

Field of application	Subject of application	Field of application	Subject of application
General	1. Knowledge base system	Management	1. Production planning
	2. Manufactoring system simulation		2. Load planning
	3. Decision-making system for manufacturing		3. Scheduling
	4. Evaluation system for CAD/CAM		4. Progress management
	5. (3 + T) dimensional CAD/CAM system		5. Quality control
	6. Enterprise automation		6. Factory management
			7. Layout of plant facilities
Design	1. Understanding of required functions		8. Project management
	2. Preparation and confirmation of specifications		9. Estimation management
	3. Recognition and understanding of specifications		10. Modelling of management
	4. Examination and determination of basic concept of design		11. Understanding of management requiring function
	5. Formation of product concept		
	6. Formation of model	Intelligent machine	1. High intelligence FMS
	7. Formation of computational and experimental procedures		2. Intelligence control machine and equipment
	8. Determination of structure		3. Learning control machine and equipment
	9. Inspection of drawing		4. Adaptive control machine and equipment
	10. Modification of drawing		5. Intelligent robot
			6. Precision and ultra-precision machining system
Production (software)	1. Understanding of required functions for production		7. Intelligent machine for specific purpose
	2. Formation of production model		
	3. Process planning		
	4. Operation planning		
	5. Intelligent NC programming		
	6. Robot programming		
	7. Software system for assembly		
	8. Software system for inspection		
	9. Software system for diagnosis		
	10. Software system for maintenance		

facturing systems. Knowledge acquisition deals with the extraction and reconstruction of specialized knowledge of experienced experts. The man–machine interface is concerned with the interaction between the expert system and the operator. Research on the items given above is required to develop the most appropriate expert system for particular applications in manufacturing.

5. ROLE OF KNOWLEDGE IN DESIGN

Information processing in tasks from design to manufacturing owes much to the intellectual work of engineers. Figure 4 shows schematically the relation between the individual tasks in the design and manufacturing processes and the levels of the intellectual work. In Fig. 4, routine work is that which has a clarified structure of the problem and can be processed routinely with the aid of the present information processing technique. Creative work, on the other hand, is that which needs the creative power and thinking of a human being and can be processed only by trained engineers. Intelligent work is classified in-between creative and routine work. At present, intelligent work is processed by the engineers; however, it has the potential possibility of being processed with the aid of a computer, once the thinking processes of engineers are analysed and formed into structures.

It is understood that routine work in the designing process is rather sparse and is limited to specific tasks. It is also to be noted that intelligent work covers a wide range of tasks from design to manufacturing. This means that there is much possibility for computer work to take over various tasks from design to manufacturing in the total manufacturing system. Expectations have been growing in recent years concerning artificial intelligence and knowledge engineering, which have the possibility of processing such intelligent work with the aid of a computer, as mentioned in the foregoing section.

As an example of the application of knowledge engineering in the early stages of design, the role of

DESIGN PROCESS		CREATIVE WORK	INTELLIGENT WORK	ROUTINE WORK
REQUIRED SPECIFICATION	UNDERSTANDING OF REQUIRED FUNCTIONS		▨	
	REPRESENTATION AND UNDERSTANDING OF SPECIFICATIONS	▨		
CONCEPTUAL (DEVELOPMENTAL) DESIGN	RECOGNITION AND UNDERSTANDING OF SPECIFICATIONS		▨	
	EXAMINATION AND DETERMINATION OF BASIC CONCEPT OF DESIGN	▨		
	CONSTRUCTION OF PRODUCT CONCEPT (CONCEPTUAL SIMULATION)	▨		
	FORMATION OF MODEL	▨	▨	
FUNDAMENTAL DESIGN	COMPUTATION & EXPERIMENT / FORMATION OF COMPUTATIONAL AND EXPERIMENTAL METHOD	▨	▨	
	COMPUTATION & EXPERIMENT		▨	▨
	DETERMINATION OF STRUCTURE	▨		▨
DETAILED DESIGN	DETERMINATION OF DETAIL STRUCTURE		▨	▨
DRAWING	DRAWING			▨
	INSPECTION OF DRAWING		▨	
PRODUCTION DESIGN	PLANNING AND DETERMINATION OF PROCESSES	▨	▨	
	PLANNING AND DETERMINATION OF OPERATION		▨	▨
COMMAND TO MACHINES	POST PROCESSING		▨	▨

Fig. 4. Major information processing in design and manufacturing and their levels of intellectual works.

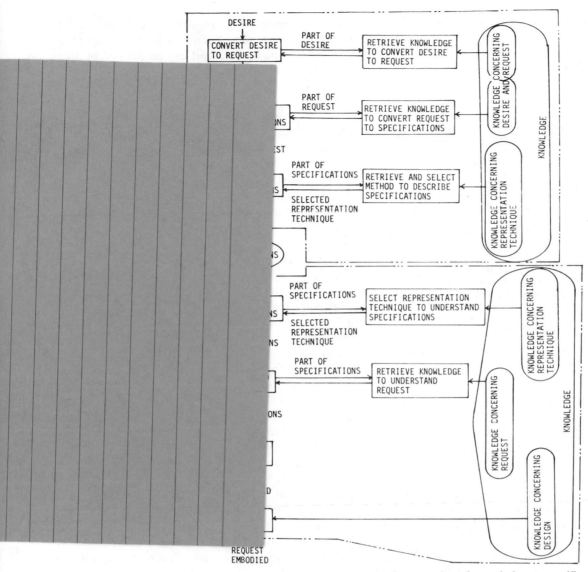

Fig. 5. Relation between activities and knowledge in conversion of user desires to specifications.

knowledge in representation and understanding of the specifications is considered here. When a product is to be designed, the desires of the user must be represented in the form of the specifications according to which the design process is carried out. The designer then understands the specifications and embodies the original desires in order to proceed to the actual design process. Figure 5 shows schematically the relations between the activities through which the desires are communicated and the related knowledge.

The desires are converted to requests, detailed, and described as specifications in the first stage where the knowledge concerning the desires, the requests, and the representation techniques is referred to. In the second stage, the specifications are read, understood, confirmed, and embodied. The

knowledge concerning the representation technique, the requests, and the design is referred to at this stage. Table 2 summarizes the major technologies to be developed to convert the desires to the specifications.

6. ROLE OF KNOWLEDGE IN PROCESS PLANNING

Process planning is recognized as the key technology to integrate the design and the manufacturing. Design information is converted to manufacturing information in this process, and computerized processing is emerging as one of the vital key technologies. Design information provided either by drawings or numerical data with technological information is first understood and manufacturing information is generated as the output of process

Table 2. Technologies required to convert requests to specifications

(1) Technology to select and extract the most influential desires and requests on specifications from the given desires and requests.
(2) Technology to retrieve knowledge concerning the extracted desires and requests and knowledge data base for this purpose.
(3) Technology to select, collect, and arrange the most important and necessary knowledge, desires, and requests referring to past experience.
(4) Technology to select the most appropriate representation method according to the requests.
(5) Technology to describe the requests by the representation method selected.
(6) Technology to judge differences between the specifications represented and the true requests of users.
(7) Technology to add a description concerning the representation method to specifications.
(8) Technology to select a common representation method between the user and person to understand specifications.
(9) Technology to select correct expressions required for retrieval of knowledge according to some of the specifications.
(10) Technology to retrieve correct knowledge concerned, according to some of the expressions.
(11) Technology to collect and arrange common expressions and knowledge with users from some of the expressions and related knowledge.
(12) Technology to let users confirm that the specifications are understood.

planning, which includes information concerning the manufacturing methods, their sequences, machine tools, jigs, etc.

So far, three kinds of approach have been taken towards this problem. The first method, known as the decision table method, is based on past experience rearranged by group technology and retrieved by referring to decision tables. This method is quite practical, though the application is limited and expansion to new products is restricted. The second method, referred to as the generative method, is based on theoretical procedures for extraction of the manufacturing information from the design information given and correlation of it to possible manufac-

turing methods. This approach has been studied mostly by academics, but there is still a gap to be crossed so that this method can be applied in practice. The third method is called the semi-generative method. It is basically similar to the generative method, but it utilizes the experience, know-how, and knowledge accumulated by process planners and designers through their many years of activities. The application of knowledge engineering is crucial to this approach.

Figure 6 shows the relation between the process planning software system and the expert system. When a problem which needs intellectural processing arises in the process planning, the problem is analysed and reasoning is carried out referring to the knowledge base. The result of reasoning is fed back to process planning. It is understood that such technologies as knowledge representation in the knowledge base, reasoning, and knowledge acquisition are quite important in developing this type of process planning.

An example of a process planning system utilizing knowledge engineering is shown in Fig. 7.[4] In this case, knowledge concerning process planning is represented by production rules. The rules are checked grammatically, converted into internal codes of the computer, examined for any mutual contradiction, and then stored in the knowledge base. The decision-making procedure utilizing the knowledge comprises the following four steps:
(1) evaluation of the condition part of the rules to determine the rules applicable;
(2) selection of a suitable rule from the applicable ones;
(3) application of the action part of the selected rule;
(4) examination of satisfaction of the terminal condition.

Application of the process planning system has shown that a suitable sequence of machine tools is

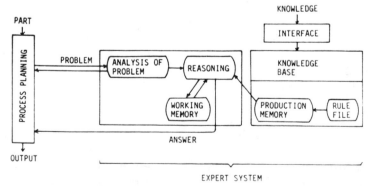

Fig. 6. Linkage of expert system to process planning.

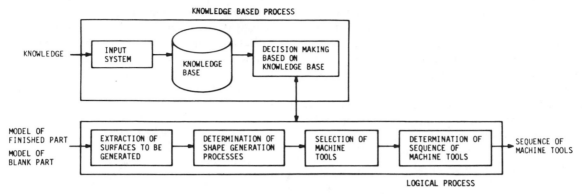

Fig. 7. Basic constitution of knowledge-based process planning system.

determined, taking into account the constraints on the machining sequence which are given as knowledge to the system.

7. HIGH INTELLIGENCE FMS

The environments surrounding the manufacturing technologies are changing, as was mentioned before. The future direction of development of the manufacturing system is considered here, taking into account the changes in the manufacturing environments. There are several possible directions for future developments of a FMS,[2] such as:

- large scale FMS which includes a flexible assembly system as well as a manufacturing system;
- cell type FMS or FMC which requires less investment and yet satisfies the minimum requirements of FMS; and
- diverted FMS to industries other than conventional manufacturing, such as food industries, etc.

The policies taken in the past to construct and utilize a FMS, such as standardization, synchronization, centralization, maximization, concentration, specialization, etc., are expected to change, reflecting the changes in the requirements of the user in the future. The requirements will be directed towards a distributed factory of proper size, operating for 24 hr a day and meeting the demands of order production. Such a manufacturing system would be highly automated and systematized. Unattended operation and high reliability are essential, and it should have a high accuracy. When one considers the realization of such a system and refers to Fig. 3 and Table 1, it is obvious that knowledge engineering plays an essential role in the realization of both the hardware and the software of such a system. Therefore, such a system would be called a high intelligence flexible manufacturing system.

8. INTEGRATED SOFTWARE SYSTEM OF ENTERPRISE AUTOMATION (EA)

After all the analyses, examinations, and discussions concerning the future development of manufacturing technology, a new concept is proposed which deals with the integrated software system of the entire enterprise; it is called enterprise automation (EA). EA consists of software systems and is illustrated schematically in Fig. 8.

The desired specifications are embodied through the various stages of design, and the product is manufactured. In the manufacturing stage, the soft-

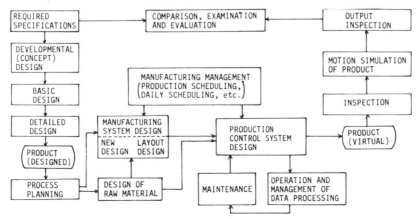

Fig. 8. Concept of EA (enterprise automation).

ware systems concerning the manufacturing system design, the production control system design, and the manufacturing management are utilized in such a way as to optimize the total processes. The product is inspected and its operations are simulated on the computer. The performance of the product is analysed, compared with the desired specifications, and evaluated. The result of evaluation is fed back to the designing stage and the whole process is repeated.

The EA software system aids in determining whether a specific product should be manufactured and how it should be manufactured.

9. CONCLUSIONS

The possible future developments of the manufacturing technologies have been examined from various viewpoints and the subjects for development in the future have been discussed. Emphasis has been given particularly to the roles of artificial intelligence and knowledge engineering in manufacturing technology of the future. Knowledge engineering itself is rather new and it is expected to have a wide range of application. The continuation of effort is essential for the future development of manufacturing systems which make the most of knowledge engineering.

REFERENCES

1. Bureau of National Life, Economic Planning Agency, Japan: *Image of National Life in the 21st Century*. 1979.
2. Ito, Y., Iwata, K.: *Flexible Manufacturing System*. Tokyo, Nikkan Kogyo Shinbun. 1984.
3. Iwata, K.: Artificial intelligence and CAD/CAM. *J. Japan Soc. Prec. Engng.* 49(10): 1448–1449, 1983.
4. Iwata, K., Sugimura, N.: A knowledge based computer aided process planning system for machine parts. *Proceedings of the 16th CIRP International Seminar on Manufacturing Systems*. Tokyo. 1984. pp. 83–92.
5. Toffler, A.: *The Third Wave*. New York, William Morrow. 1980.

CHAPTER 2

EXPERT SYSTEMS FOR COMPUTER AIDED DESIGN

Presented at the CASA/SME AUTOFACT 5 Conference, November 1985

Expert System Model of the Design Process

by Faith Kinoglu
Don Riley
and
M. Donath
Control Data Corporation

INTRODUCTION

Artificial Intelligence (AI) is the branch of computer science that uses computers to reproduce behavior usually associated with human intelligence. Thus, AI is devoted to studying, developing, and applying computational approaches to intelligent processes (8). Significant branches of AI include computer vision, natural-language understanding, expert systems, knowledge based systems, computer-aided learning, and speech recognition and synthesis.

For years AI has been a research area carried out mostly by a few major universities and research labs. However, in recent months, AI and Expert Systems have received tremendous publicity. The decline of hardware costs and the availability of reliable and supported software tools have enabled AI to emerge as a realistic and practical technique that offers a large variety of useful applications in manufacturing and engineering areas.

Expert systems produce (3,8,9) intelligent behavior by operating on the knowledge of a human expert in a well-defined application domain. The ability to operate on this knowledge gives the expert system the capability to perform its task at a skill level usually associated with the expert. Because knowledge is the key ingredient in an expert system, such systems are often called knowledge-based systems.

These systems include a knowledge base containing facts, rules, heuristics, and situation patterns, and an inference system that makes decisions within a domain. Recent applications of expert systems have demonstrated the potential to achieve a high level of human performance while preserving knowledge that otherwise might be lost by attrition, retirement, or death. These systems also have demonstrated the ability to improve upon the performance of average individuals by providing them access to the encoded knowledge of the scarce experts.

The basic structure of an expert system is illustrated in Figure 1. The components include: input/output facilities, that allow the user to communicate (sometimes referred to as Natural Language Interface) with the system and to create and use a data base for the specific case at hand; an inference engine, that incorporates reasoning methods, which in turn act upon input data (i.e., rule interpreter) and knowledge from the knowledge base, to both solve the stated problem and produce an explanation for the solution; a knowledge base, and (perhaps) a knowledge acquisition facility, which allows the system to acquire further knowledge about the domain from experts and/or from libraries and data bases. The inference engine acts as the executive that runs the expert system. It fires rules according to a built-in reasoning protocol, and by so doing performs actions that lead to solution of the problem.

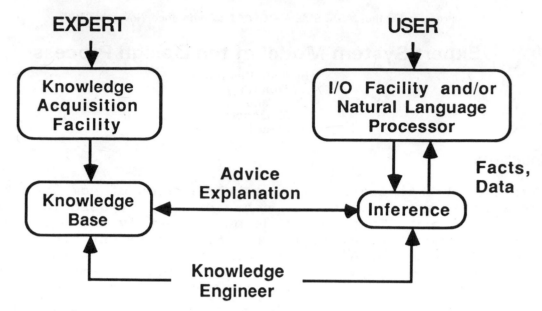

Figure 1. Structure of an Expert System
(Dym, Clyde L., Engineering with Computers, V.I, 1985)

At the same time, the inference engine may change the knowledge base by adding new knowledge to it. The knowledge base contains the domain-specific knowledge which provides the context for the specific applications of the expert system.

The expert system (or knowledge-based systems) tasks are: interpretation, the examination of data to determine their meaning; diagnosis, the process of error/fault identification mechanism; planning, the preparation of an agenda of activities to achieve a set of goals and/or tasks given a set of possibly changing constraints; monitoring, the continuous interpretation of signals; prediction, the forecasting of future performance from existing models; and design, the definition of requirements to create new objects. Some current expert systems application areas are: medical diagnosis; equipment repair; computer configuration; image synthesis and analysis; signal interpretation and mineral exploration. Recently, however, a number of researchers and large corporations have begun to develop domains in which the fundamental task is design and/or manufacturing process planning.

The tasks of interpretation, diagnosis, and monitoring are especially suitable for engineering and manufacturing problems. Since each of these tasks deals with the ability to obtain data, make judgments (or reason), and take corrective action, they will be very suitable for the following tasks:

- monitoring and diagnosis for a manufacturing operation (i.e., real time process control)
- failure detection
- self-diagnostic capabilities for maintaining automated equipment
- on-line evaluation of non-destructive test results
- monitoring and controlling machine utilization

In many of these applications the expert system needs to be linked to some sensors and signal processors (microprocessors) to provide the necessary data about the current state for a particular task (monitoring), which then can be interpreted and diagnosed by rules provided in the expert system.

The tasks of planning and design can be applied to engineering and manufacturing areas since these functions are central to the product design. Artificial Intelligence has applications in virtually all major areas of manufacturing, especially those that rely on experience and know-how (e.g., process planning). Some potential tasks could include the following areas:

- design of mold, dies, and runners in casting and injection molding

- selection of tool materials, feeds, and speeds of cut in metal processing

- generative process planning

- assembly planning

- factory management

- cost estimation and budgeting

Design tasks have characteristics (4, 5, 6) that make them appropriate to address with expert systems techniques. This paper describes a prototype knowledge-based system that provides an intelligent redesign mechanism for the product design (from conceptual work to the manufacturing process creation).

THE DESIGN PROCESS

As illustrated in Figure 2, the design process starts with a definition of a need, which can be satisfied by the product. This need may take many people and much time to define, or may be specified by one person in a few minutes. The definition is then gelled by formation of a general concept of the solution required, which gives rise to a specification for the product. As a result, the various phases of the product design process interact with each other. However, design is not a simple serial process since each step depends on the result of the previous one.

A design project is initiated with a set of requirements generated by the sales/marketing force for the product design group (engineering and manufacturing). These requirements are projected into ideas about things; which in turn are translated into engineering prescriptions for transforming suitable resources into useful, physical objects. However, the product requirements, which are based on some need, are usually not well-defined. For example, the requirement can be specified as "a need for a system that produces rotational motion at about 10 hp and 800 rpm which can be used in a conveyor belt mechanism." The design process begins with the conceptual work, continues with feasibility study and preliminary design activities, and finalizes with detailed design and manufacturing engineering processes which are used to prepare the production

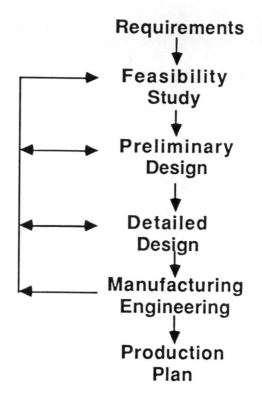

Requirements

Feasibility Study

Preliminary Design

Detailed Design

Manufacturing Engineering

Production Plan

Figure 2. Phases of a Product Design

plan. However, each one of the design stages are made of multiple steps that interact with each other.

The first step in the design activity is the feasibility study; the purpose is to achieve a set of useful solutions to the design problem. It establishes whether the problem posed by the design project is solvable and whether there are solutions that are likely to be useful.

The preliminary design phase starts with the set of useful solutions which were developed in the feasibility study. During this stage the best applicable design alternative is selected and major design parameters are evaluated for various types of analyses. The preliminary design is intended to establish an overall concept for the project which will serve as a guide for the detailed design.

The detailed design begins with the design concept evolved in the previous step. Components and subsystems, partial prototypes, and complete prototypes are tested as the need for information arises. With the design concept and the results from preliminary design, an overall synthesis is accomplished (i.e., master layout), and the detailed specification of components is carried forward. The choice of material for the individual parts is defined; for example, heat treatment and surface treatment may be prescribed if certain specifications require it. The producibility of the parts must be considered and, at least in a general way, the production process established for manufacture. The detailed drawings are prepared and reviewed, and all the necessary descriptions (such as specifications, dimensional and form tolerances, assembly instructions, standard notes and symbols, special sketches, and revisions) are made available for manufacturing.

During the manufacturing engineering stage the following tasks are performed:

- detailed planning of the manufacturing processes (process planning). This step is particularly important, because design features that will cause problems in the production may be discovered.

- design of tools and fixtures necessary to produce (this is based on the previous step, i.e., process plan).

- planning and specifying the quality control system and standard times.

Finally, in the production planning phase the factory management techniques will be applied to the final design in order to produce the design, together with the planning for distribution, consumption, and retirement of the product.

The product design process is a highly iterative process. Each stage is interconnected and is interdependent upon other systems. As Asimow (2) put it:

"The design process describes the gathering, handling, and creative organizing of information relevant to the problem situation; it prescribes the derivation of decisions which are optimized, communicated, and tested or otherwise evaluated; it has an iterative character, for often, in the doing, new information becomes available or new insights are gained which require the repetition of earlier operations. Some of the operations are qualitatively logical in character, like reasoning from verbal propositions; some are based on subjective evaluations, as in comparing or combining unlike values; many are amenable to quantitative analysis and to computer applications, as in optimizing an analytically formulated representation of a problem solution. For the most part the techniques associated with each operation in the design process are of such great generality that their usefulness is not limited to any particular step."

The availability of CAD/CIM has changed the design process (Figure 3) by providing the designers computer assisted design tools for generating the new designs from scratch and/or by modifying an existing similar design. The existing designs can be located and modified during the preliminary design stage where the designer can make the necessary changes to the old design in order to accommodate the new requirements. The new design can then be analyzed mathematically for various conditions (load, stress, etc.) before the producibility check. During the manufacturing process design, like in engineering design, an existing similar process plan can be accessed and modified for the new design. The engineering and manufacturing data related to the same design are kept together by using group technology techniques which integrate the product design process.

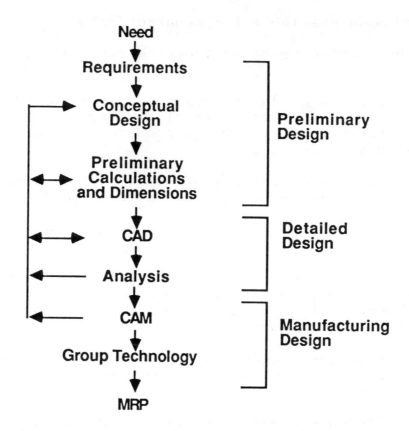

Figure 3. Product Design Process in Light of CAD/CIM

THE PROTOTYPE MODEL

The knowledge based system that is currently under development aids the designer during the product design process starting from the conceptual design stage. The objective of the prototype model is to emulate the engineering and manufacturing design processes through intelligent redesign.

The subsystems of the prototype (Figure 4) include:

- input/output facilities, that allow the user to communicate with the system through various windows.

- a user dictionary, which contains the user defined vocabulary (e.g., abbreviations, personal definitions) in order to inter-face the global knowledge base. The input/output facility also includes the tools to update/create the user dictionary (in real-time). Simple parsing routines are used to facilitate the interface.

- a local knowledge base, which contains knowledge that is ex-clusively used by individual users. Once the local knowledge is widely accepted, it can then be transferred to the global knowledge base. The local knowledge base acts like a scrap book for the designer during the preliminary design process by

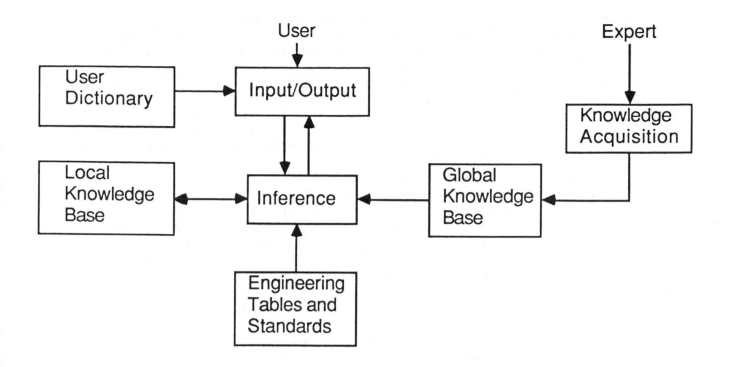

Figure 4. The Subsystems of the Prototype

allowing him/her to define new rules and procedures, and apply
them in real-time. The new rules and procedures are then
checked against the global knowledge for constraint violations.

- an inference engine, which incorporates reasoning methods to
  act upon user requests and knowledge from the global and local
  knowledge bases by accessing the CAD/CIM data bases, and
  engineering tables and standards. The CAD/CIM data base gives
  access to the existing designs in the system which allows the
  user to locate any similar parts. The engineering tables and
  standards contain some of the characteristics that can be found
  in various engineering handbooks (e.g., material data, design
  codes, etc.).

- the global knowledge base, which contains the knowledge necessary
  to design specific components, including facts, beliefs, and
  heuristics to make decisions. The knowledge is represented
  in frames to expose the attributes of individual concepts,
  production rules to manipulate the data attributes, and a
  semantic network to relate concepts to each other.

- a knowledge acquisition facility, which allows the system to
  acquire further knowledge from the experts. The facility allows
  any expert to enter the new knowledge through "knowledge
  windows."

The process (Figure 5) begins by the user querying the system with a design concept in mind. This is done by specifying the application area(s) and desired function(s) of the new design. The information is then translated through the user dictionary in a query form to the global knowledge base. Once the matching existing design is located, then the system will use the knowledge base (i.e., how to design and design procedures) to perform preliminary and detailed design calculations using the engineering tables and standards. The user is also allowed to re-specify the criteria and procedures for the new design.

During the preliminary and detailed design processes the user can introduce new procedures and variables and apply them to find out the effects on the new design. The system will record the new procedures into the local knowledge base by associating them with the design and global knowledge base. At the end of the design process the system will generate a partial classification code (including major shape and functionality features) which then can be used to search the group technology data base for a similar process plan.

The manufacturing process (Figure 6) starts with the existence of a matching process plan and continues with the semi-automatic update of the old plan for the new design. The system has production rules associated with various class code selections which are used to compare the old and new designs to make the necessary changes. However, the user is asked to verify the final process plan before it is stored in the data base.

The prototype was developed using TLC-Lisp (1) on a Zenith microcomputer (Z-150 and Z-100) configured with 10 megabyte hard disk and 702K memory. Currently, the prototype model is being translated into Common Lisp (7) in order to continue the development effort on a Control Data Cyber NOS/VE system.

The prototype model uses an event driven search technique which employs numeric procedures and engineering standards when reasoning in the presence of uncertainty. However, the prototype allows users to accept or override the decision made by the control structure. In addition, a shallow justification and explanation facility is provided to the user to explain the reasoning.

The user interface is done through windows that display the internal reasonings and numerical calculations as they progress (Figure 7). Using the windows the user can follow the process and interrupt the process to try other alternatives by changing the constraints and/or processes, and to query the system about its reasoning. The process that is interrupted can later by continued by a warm start.

In the prototype model each product is represented by a frame structure (Figure 8). In addition, each frame structure is made of various data frames, where every data frame represents the most general level of the solution containing the conceptual store of variable names and values. These variables may be from many domains expressed numerically (e.g., dimensions) and/or by words (e.g.,

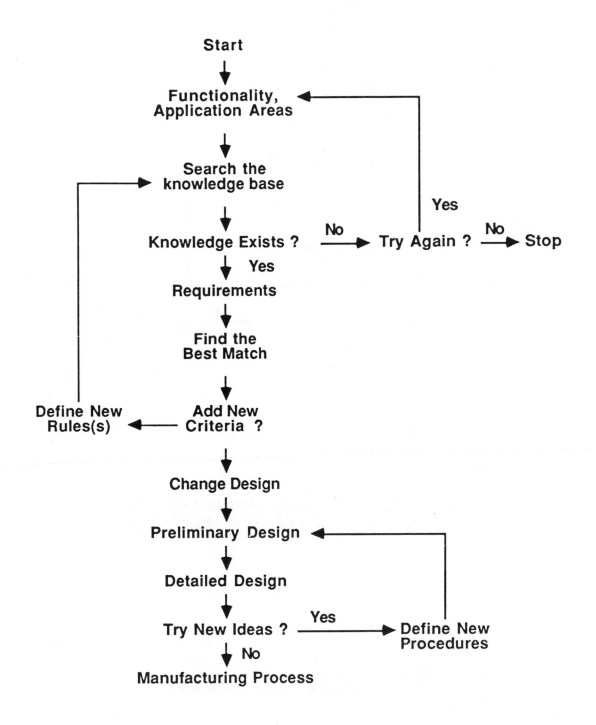

Figure 5.   Engineering Design Process

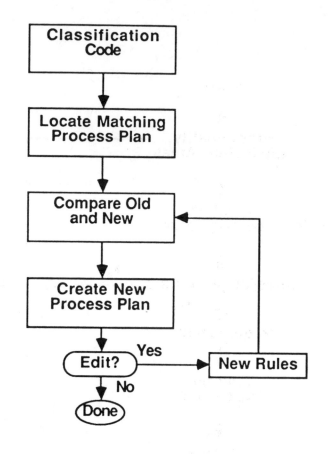

Figure 6.   Manufacturing Process

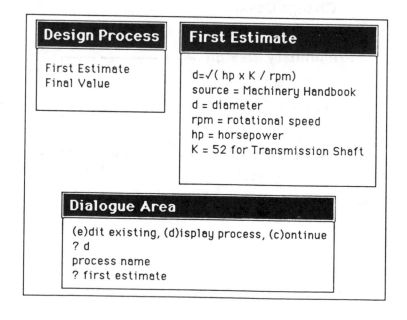

Figure 7.   An Example Window

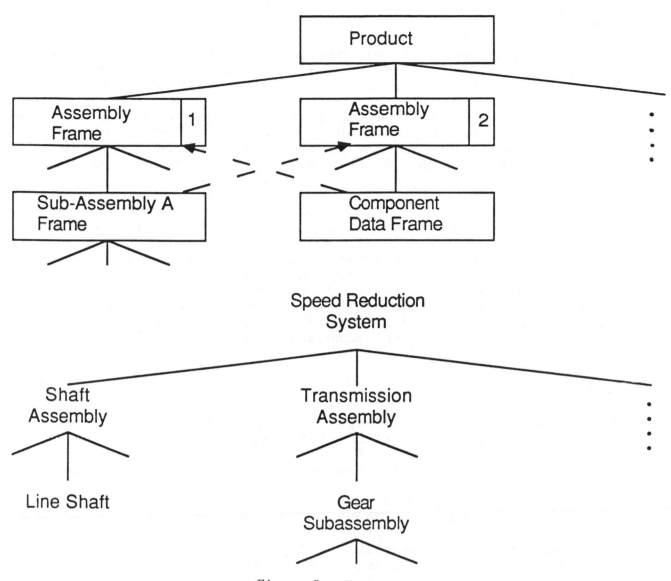

Figure 8. Frame Structure

vendor name). For example, the frame structure for a speed reduction assembly may contain the ball bearing as one of the data frames. Furthermore, the ball bearing data frame can contain the necessary design variables such as load capacity, life, and supplier name.

The prototype model employs constraint propagation in order to establish the interactions (the relations) between various stages of the design process. The design constraints are applied to each of the data frames. These constraints are in the form of variables that are instantiated by other data frames through influence rules that link various data frames. For example, in the speed reduction assembly, the shaft diameter value will be shared by ball bearing and seal data frames.

The future work will include the integration of the knowledge base with the CAD/CIM Data Base in order to control the design changes more effectively, interfacing with external data bases, and expanding the knowledge acquisition facility.

CONCLUSION

The prototype model was developed in an attempt to automate the design process intelligently by networking multiple domains. The data frames coupled with production rules which are linked by a top-down constraint propagations worked well in modelling the design process. During the development of the model, a considerable amount of work was spent in resolving such issues as: when to translate the local knowledge into the global knowledge base; the level of interruptions that should be allowed to the user; and strategies in instances of conflicting reasoning.

The prototype model has demonstrated that artificial intelligence and expert systems tools are very valuable tools for engineering and manufacturing design processes. Although the development of an expert system requires a substantial amount of resources, the payback in the near future will be high enough to attract the various engineering and manufacturing disciplines. Already various universities and large corporations are investigating the potential applications of AI, and developing expert systems to model various tasks from engineering and manufacturing areas.

REFERENCES

1. Allen, J., 'TLC-Lisp Reference Manual,' The Lisp Company, Redwood Estates, California, 1984

2. Asimow, M., 'Introduction to Design,' Prentice Hall Inc., 1962

3. Davis, R., 'Expert Systems: Where Are We? And Where Do We Go From Here?', AI Magazine, Summer 1982

4. Freeman, P. and Newell, A., 'A Model for Functional Reasoning in Design,' Int. Joint Conf. on Artificial Intelligence, 1971

5. Mastow, J., 'Toward Better Models of the Design Process,' AI Magazine, Spring 1985

6. McDermott, J., 'Domain Knowledge and the Design Process,' ACM/IEEE 18th Design Automation Conference, 1981

7. Steele, G. L. 'Common Lisp,' Digital Printing, 1985

8. Winston, P., 'Artificial Intelligence,' Addison Wesley, 1977

9. Winston, P. and Horn, B., 'Lisp,' Addison Wesley, 1981

Presented at the CASA/SME AUTOFACT 6 Conference, October 1984

Automating Design Invention

by Michael G. Dyer
and
Margot Flowers
University of California at Los Angeles

1. Introduction

In manufacturing, most current "robots" are not really robots in the sense of being intelligent autonomous agents, but rather are complex manipulators going through preprogrammed motions, or at best, being guided by primitive pattern recognition systems.

Thus, current interest in applying Artificial Intelligence (AI) technology to manufacturing engineering has focussed on the use of intelligent vision systems to control and direct robot manipulators for automated factories. Clearly, a robot capable of recognizing and manipulating images will be able to grasp randomly oriented objects, or keep from bumping into obstacles. But besides these immediate tasks, what might a robot do while it's *not* working (heaven forbid) on the task of directly recognizing images or manipulating those objects it sees before it?

As AI researchers we are interested in understanding all of those cognitive skills which engineers possess, and in this paper we address the design (rather than manufacturing) side of mechanical engineering. Specifically, we ask the questions:

> Could a robot have imagination?
>
> Can a computer be programmed to invent
> or design novel mechanical devices?

Computers already aid humans in the design process. For instance, in textile design, simple graphics programs can generate myriads of patterns from which a human can select something attractive. But we are not interested here in cases where the computer simply serves as a tool for a human designer. In this paper we address the queston of whether, and to what extent, processes of creative thought can be encoded into a computer so that the computer can be said to have "invented" something "on its own".

We will first briefly describe pioneering AI research in formalizing heuristic "rules of creativity" in the areas of mathematics and science. Then we will show how this approach could be used in mechanical design.

2. AI Research in Creativity and Invention

Until this last decade, the areas of creativity, discovery, and invention had always been favored by sceptics as areas in which no inroads had ever been accomplished by AI researchers, and as areas in which success would never be obtainable. However, the pioneering work of Lenat (1977), in the area of mathematical discovery, showed how a level of creative behavior could be programmed into a computer.

2.1. AM

Lenat's program, called AM, attempted to model processes of concept invention and discovery. AM started out with about a hundred elementary concepts from set theory, such as set membership, set union, the empty set, and so on. Each concept consisted of its name, a definition of the concept, some examples of the concept, pointers to other concepts which were generalizations of it, pointers to more specialized concepts, some working hypotheses to be explored concerning the concept, and finally, a measure of how interesting or worthwhile AM thought this particular concept was.

In addition to these primitive concepts, AM also contained about 250 heuristics, or "rules of thumb", for 1) filling in missing information about an existing concept, 2) creating new concepts from old concepts, and 3) determining the current worth of a new concept, or reevaluating the inherent interest in an already existing concept.

AM operated by taking the concept with the highest interest from an agenda, or "task list", of concepts. AM then applied its heuristics in an attempt to either fill in missing information or invent a new concept. As new concepts were created, AM calculated their worth and placed them into the agenda according to how interesting AM felt the new concepts were.

Since AM began with set-theoretic concepts, Lenat expected AM to further explore concepts in set theory. However, early on AM invented the notion of "number" by applying the concept of set length to sets consisting of identical items. As a result, AM's explorations took it off into number theory. By applying the known concept of set union to the concept of number (i.e. the length or cardinality of a set), AM arrived at the new (for it, at least) concept of "addition". Similarly, by using the notion of set difference, AM discovered the concept of "subtraction".

After inventing a new concept, AM would stop and ask the user if he wanted to name the new concept. In most cases, concepts invented by AM had already been discovered by mathematicians long ago. In such instances, Lenat selected names which corresponded to those already used by mathematicians.

While rediscovering addition and subtraction, AM also reinvented concepts, such as the identity concept for addition. That is, AM noticed that the length of the empty set (i.e. the "number" zero) added to any set of length n yielded a set of identical length.

For instance, one of AM's heuristics was:

> If you have invented a concept C by using an operation F
> Then you can create a new concept C' by applying the inverse of F.

By applying this heuristic, AM rediscovered the concepts of "multiplication" and "division". AM then invented the concept of "even numbers" by noticing the results of dividing various numbers by the number two. Dividing by 1 also lead to the identity concept for division. After about 1 cpu hour of running time, AM had reinvented both the notion of "prime numbers" and that of Goldbach's Conjecture -- i.e. that every even number greater than 2 is the sum of two prime numbers.

It is important to realize that AM did not know or care whether Goldbach's conjecture is true. That is, AM was not capable of proving any theorems in mathematics. AM had not been designed to do so. Instead, AM was designed to create new concepts by applying heuristic "rules of invention" to existing concepts in order to create novel concepts.

Other notable concepts rediscovered by AM included: odd numbers, squares, square roots, exponentiation, and perfect squares. AM also discovered a new concept not really noticed previously by mathematicians -- i.e. the concept of maximally-divisible numbers, or numbers with an abnormally high number of divisors.

In addition to its rules or heuristics of invention, AM also contained rules for determining the "interest" of a concept. For example, here is one rule of interestingness:

> Increase the inherent value of a concept if it has been reinvented or rediscovered by the application of more than one rule.

For instance, since division was arrived at by both applying the inverse of multiplication and by the use of successive subtraction, it was deemed a very interesting concept and was therefore worked on more than other concepts.

In AM's best run, it discovered 25 winning concepts, about 100 acceptable concepts, and had wasted its time working on about 60 other concepts which turned out to be worthless.

Unfortunately, after an hour's run, AM failed to invent more novel concepts. This occurred because AM lacked the ability to synthesize new rules of creativity from old rules. AM also lacked the necessary knowledge of mathematics needed to go on. However, in general AM showed that it is possible to mechanize aspects of creativity in order to produce a system capable of inventing new ideas from old ones. AM also demonstrated that such a system could, to a certain extent, decide on its own what concepts were worth pursuing.

Lenat also briefly applied AM to the domain of geometry. He gave AM the primitive concepts of point, line, angle, triangle and equality of points etc. Applying its heurstics, for example, AM rediscovered the notion of "congruence" and similarity of angles.

One interesting avenue of research is that of applying AM to itself in order to see if it is possible to "bootstrap" creativity, thus producing a system which can continually invent new heuristics for creativity (Lenat 1983).

2.2. BACON

Since the work of Lenat, others have applied his approach to automating processes of creative invention and discovery in the areas of the physical sciences. For instance, the BACON program (Langley et al. 1981) has rediscovered the ideal gas law , Coulomb's law, Kepler's third law of planetary motion, Ohm's laws, and Galileo's law of constant acceleration. BACON is given data from the physical sciences and then uses various heuristics to postulate laws which explain the data.

Here is a simplified example of a rule used by BACON:

> If a1 is an independent variable, and a2 is a numeric
> dependent variable, and the values of a1 change
> when the values of a2 change,
> Then propose an intrinsic property whose values
> are taken from the values of a2.

For example, this rule is used for postulating an intrinsic property of a nominal variable, such as conductance.

3. The Door to Engineering Design Creativity

In this paper we show how rules of invention might be used in the area of design. In doing so, we relied on informal introspective protocols. That is, we have tried to be inventive ourselves and then have attempted to introspect on what we seemed to be doing. Specifically, we tried to invent a novel version of a familiar mechanical device -- i.e. the door.

We chose the common door as a starting point for the following reasons:

1. It is a relatively simple mechanical object, yet complex enough to be interesting.

2. It is an extremely common class of items. There are many different types of doors which have already been invented. As a result, we will be able to see at least whether or not our rules of invention are capable of reinventing useful doors. This allows us to see the range of invention capable of our hypothetical system.

3. Doors are relatively easy to draw, and since people already know what doors look like, little effort need be expended in detailed drawings or descriptions of any invented objects. Furthermore, there should be almost immediate consensus on the worthiness of a specific door.

4. By choosing an everyday item, we can show what's required for those generic processes of invention which must be understood and formalized before sophisticated engineering design can be automated.

The point of our exercise was not necessarily to come up with a really novel door (i.e. one worth patenting), but to show the reader the feasibility of design invention in engineering using AI techniques, and where the problems and difficult issues arise. By end of this paper, readers will appreciate why we did not select motors, bicycles, machine tools or other more complex objects.

In the following sections we show how heuristic rules of creativity can be applied to reinvent various door types, starting from an initial, prototypic door, shown in figure 1 below.

Figure 1:   Common 2-hinge door
            with handle.

In the next sections, we will propose and apply general discovery heuristics, or "rules of creativity" (c-rules). In going through this exercise, we will touch upon the problems of:

- Combinatoric Search
- Naive Physics Knowledge
- Problem Solving on Constraints
- Episodic Memory and Analogical Reasoning

3.1. Discovery and Invention Through Combinatoric Search

The process of invention can viewed as a search through a large space of states representing objects or ideas (Pearl 1984), (Nilsson 1980) . Movement from one state to another is accomplished by the application of an operator. In our case, the state space consists of object designs and the operators are "creativity rules" or rule schemas which modify or combine existing objects to yield new designs. Consider the rule below:

> C-rule 3.1a: (Feature Variation)
> To invent a new object O' from O, select a feature
> F of O, select a scale S associated with F, and
> vary feature F along scale S.

This is a very general rule, and is capable of creating many new objects. For instance, we can select shape as a feature and alter that. If our intelligent design invention system (IDIS) knows about different shapes, then C-rule 3.1a can be used to generate circular doors, triangular doors, hexagonal doors, and so on.

Another feature is hinge location. If our IDIS selects hinges and varies their placement along the circumference of the door, it might invent an "upright trap door" in figure 2.

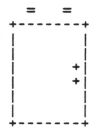

Figure 2: Upright trap door

If width or height is selected as a feature and varied, then very wide or very thin doors would result.

> C-rule 3.1b: (Variation Combination)
> To create a new object O' from O, apply feature
> variation to more than one feature at once.

Using C-rule 3.1b, an IDIS can now generate a combinatorially large number of new objects, such as the hexagonal upright trap door in figure 3.

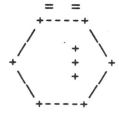

Figure 3: Upright hexagonal
trap door.

There are several issues, however, that this exercise has brought up.

1. When do we know that we have something really new?
2. How do we know that what we have invented is worthwhile?

As we can see, it is easy to create a large number of objects trhrough feature variation and combination, indeed, a combinatorially explosive number of objects. However, most of these objects are simply slight variations of the original object, and, besides maybe the trap door, are really not what most people would consider a truly novel door. For example, varying the placement of the door handle will give us a door with a handle in the middle, another with a handle toward the bottom, and so on, as in figure 4.

Most of these doors are just not very interesting. Further more, most of these doors *sim-*

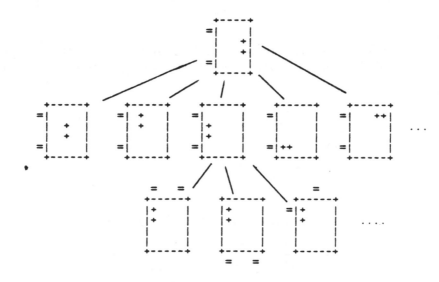

Figure 4:   Fragment search space
of door designs

ply will not work. Remember the joke about the person who proposed an idea which could not work, and when this fact was pointed out to him, replied "I did the hard creative work coming up with the idea. The rest is just minor details I leave to others to figure out."

What an IDIS needs are heuristics which eliminate minor variations from being considered as major inventions and, more importantly, a way of recognizing *constraints* or impendiments on the worth of the designs being created. Notice in figure 4 that there is a door with one hinge on either side of a corner. Clearly this door will not swing. But how would an IDIS realize this?

3.2. Naive Physics in Engineering Design

Actually, our ability to guess or predict that a door will not swing comes from our *naive physics knowledge* of materials, objects, forces and motions, such as hinges, swinging, arcs of rotation, and the like (Forbus 1981), (Hayes 1979), (deKleer 1981). Without a way of representing and applying this knowledge, constraint analysis is impossible. It may be that a constraint can be eliminated through problem-solving mechanisms, but we must first be able to recognize a problem before we can solve it. Furthermore, a solution to a problem requires reasoning about that problem, which again requires knowledge of what the problem involves.

Few people are professional engineers or physicists, yet everyone has a well-developed naive model of the physical world by the time they are a few years old. Children giggle and laugh at cartoons which are full of physical violations. For example, an elephant will go behind a spindly little tree and its entire body will disappear, or a coyote will climb a ladder into the air and keep climbing although there are no more rungs, only to fall after realizing there are no more rungs, etc.

Although engineers have very specialized, scientific knowledge and mathematical models concerning the nature of physical world, they rely heavily on common sense knowledge to determine whether or not their more specialized models are accurate. If we arrive at designs which violate our common sense notions of pressure, tension, thickness, etc., then we recheck our assumptions.

Unlike scientific knowledge, naive physical knowledge is nowhere written down, since it is considered 'obvious' by everyone. However, any intelligent automated inventor or designer must possess this 'obvious' knoweldge. However, it is the case in AI that naive physical knowledge is not yet formalized or well understood, therefore it is actually easier at this point in time to model complex equations on a computer than to model common sense notions of the physical world.

A few months ago we were both watching a 14 month old girl, named Tessa, learning to eat soup with a spoon for the first time. She was making quite a mess and at first, it seemed that what she was doing was quite random. However, in a 30 minute period we saw her do the following:

1. Put the spoon in the soup. Lift it up. Bring it toward her. Tilt it. Watch it spill on her clothes.

2. Put the spoon in the soup. Lift it up. Tilt it. Watch it spill into the bowl.

3. Vary the speed of tilting. Tilt it slightly and watch it drip. Tilt it more and watch it drip faster. Tilt it more and watch it pour as a stream.

4. Watch varying degrees of dripping and various speeds and thicknesses of the streams and drops being made.

5. Watch the splash made by the drips and streams. Hit the spoon on the soup. Watch the splashes. Vary the speed of impact and see the effect.

6. Try 1 through 5 with the other hand. Notice that effects are invariant with choice of hand.

7. Lift the full spoon over the table. Pour soup onto the table. Watch a puddle form.

8. Vary tilting and notice how the puddle grows as more soup is added.

9. Start new puddles. Add to old puddles. Notice that larger puddles grow more slowly than smaller puddles.

10. Grab a smaller spoon. Repeat 1 through 9.

These observations convinced us that Tessa's actions were far from random. In fact, they seemed extremely scientific. Consider just a few of the rules of naive physics for liquids which Tessa probably accumulated during that time period:

NP-Rule 1: Liquid stays horizontal even when its container tilts.

NP-Rule 2: Liquid spills (i.e. will leave a container which has a section below the liquid's level).

NP-Rule 3: The thickness of the stream increases with the amount of liquid leaving.

NP-Rule 4: Tilting increases the amount of liquid leaving.

NP-Rule 5: Small amounts of falling liquid form into drops.

NP-Rule 6: Increased spilling increases the closeness of the drops.

NP-Rule 7: Larger amounts of liquid form streams.

NP-Rule 8: The thickness of the stream increases with pouring.

NP-Rule 9: Liquids form puddles.

NP-Rule 10: At constant pouring, puddles grow fast at first and then slower. (It will be years before Tessa learns about inverse square or other expansion laws.)

While 'playing' with her soup, Tessa also acquired numerous invariance laws (such as the fact that the hand used is unimportant, or the spoon used is unimportant unless it is a different size). To all of us, these rules are obvious, but they are not obvious to Tessa and other infants, who must learn them. Remember playing with a hose of water flipping it and watching how the stream forms patterns in the air? Years later this knowledge may be applied by an engineer who is designing a liquid spray process.

3.3. Problem Solving in Invention

So far, we have not really invented many doors using C-rule 3.1a. However, there are many ways in which we can alter a given design object in order to produce new objects. Below is a variant of C-rule 3.1a:

C-rule 3.3a: (Component Alteration)
To invent a new object O' from O, select a component C of O and alter the number of those components.

That is, instead of just modifying a feature along a scale, increase or decrease the number of components used in making up a strutured object. A trivial example involves increasing the number of hinges on our prototypic door. This gives us the continuous hinge or multi-hinged door in figure 5.

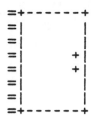

```
=+------+
=|      |
=|      |
=|    +|
=|    +|
=|      |
=|      |
=+------+
```

Figure 5: Continuous hinged door

This is not very exciting, but let us explore the use of C-rule 3.3a further. What other components can be altered in number? For example, our prototypic door has a single plane or "slab". What if we were to increase the number of separate slabs in a single door? Again, there are numberous ways of doing this. For instance, we can slice our existing slab vertically down the center. This leads us to a half door swinging on its hinges but leaves us with an unattached other half just standing there with no support or ability to open. However, we can solve both the support and rotation problem by attaching hinges to the second slab. By varying the placement of the hinges, an IDIS would reinvent both the accordian door and the bar door in figure 6.

This approach opens up (no pun intended) numerous other doors to explore. For instance, increasing the number of slabs to N simply makes an N-piece accordian door. Trying to increase the number of slabs past two for a bar door, however, requires some thought (figure 7).

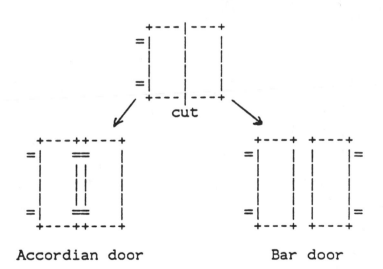

Accordian door   Bar door

Figure 6: Increasing the Number
of door planes leads to accordian
and bar doors.
```
```

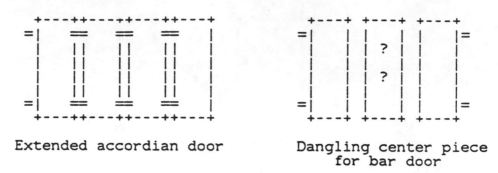

Extended accordian door

Dangling center piece
for bar door

Figure 7

There are several ways we could solve the problem of a center piece on a bar door. For one, we could hinge it at the top and make it a swinging vertical trap door, but that will make for an awkward door, i.e. one that, to open, requires moving 3 things at once in 3 different directions: the left slab rotating to the left, the right slab rotating to the right, and the middle slabe rotating upward. Another possibility is simply to use the accordian door solution approach and attach the unconnected center slab to one or the other of the wing slabs. With a little adjustment, we get both a split accordian door and a swinging accordian door, as shown in figure 8.

Figure 8 shows how each door differs in the way it can be opened. With an accordian bar door you can open each connected slab like an accordian while swinging each half separately like a bar door.

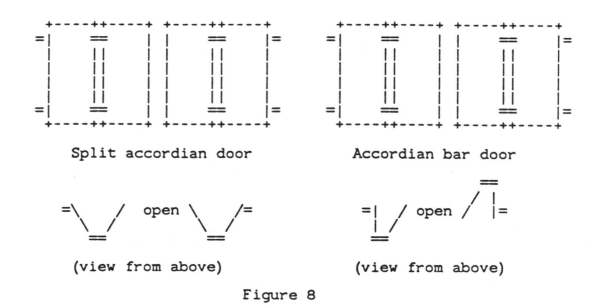

Split accordian door

Accordian bar door

(view from above)

(view from above)

Figure 8

The accordian form opens up a whole new realm of possibilities. For instance, we can increase the number of slabs by cutting and then hinging horizontally, thus producing the venetian blind door in figure 9.

From these informal protocols, the following obervations can be made:

1. Problem solving is central to invention.

We see what an important role problem-solving plays in the process of invention. In most cases, the doors we came up with all had something wrong with them, and required adjustments before they could be considered useful. For instance, we created a bar door with three slabs, and then realized that the middle slab was unattached. Thus we had to either reject that discovery path, or try to fix the problem. The inventor who gives up

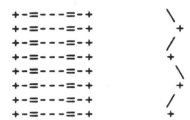

Venetian Blind door which
opens and closes vertically

Figure 9

whenever he encounters a problem may fail to explore paths that will lead to novel design objects. One the other hand, encountering problems and assessing the difficulty of their solution can lead as a way to eliminate the need of exploring worthless possibilities.

Problem solving is important as a way of pruning our space of door possibilities. Consider just a few of the different ways door slabs could be cut (figure 10).

In each of these cases, the resulting doors either do not function properly, or open awkwardly. Only bar door A will swing open, but the person pushing door A open will have to apply a long rotational arc to one side and a small arc to the other side. The way to solve this problem is to know how to apply naive physical knowledge concerning radii and rotations. This requires knowing that the door, viewed from above, is tracing out a radial arc. By using naive physical reasoning, the cut choice on bar door A can be moved until the arcs of movement are symmetrical. With problem solving ability, coming up with the perfect idea at once is not as critical. As we can see in this case, if the center point had not originally been chosen as the cut point, problem solving would result in that point ul-

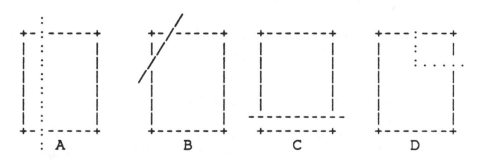

Figure 10: Cutting door slabs

timately being selected anyway.

2. Need for operational knowledge

Notice also, we could not simply say: "Let's increase the number of slabs." and presto, have it happen in just the right way. In fact, the decision to cut the slab vertically down the center required knowing first that cutting exists as a possible operation. That is, in addition to knowledge about components that make up and object, and how they go together to form the object, IDIS must also know about which operations can be performed on components. One such operation is cutting. That is, IDIS must know that cutting a plane with a single line will produce 2 planes with the same total area. This may seem obvious to us; but this 'obvious' knowledge must be encoded somewhere in an IDIS.

In some cases, operations are best associated with objects and should contain information about how the operation transforms the object. In other cases, operations are better associated with the materials used to make an object or component.

Some of this common sense knowledge is:

> Shaving an object makes its dimension shrink along the dimension perpendicular to the shaving motion.

> Drilling results in a hole with a diameter the size of the drill bit.

and so on. For instance, once an IDIS decides to try increasing the number of slabs by one, it cans use this intended result to index a method for achieving it, and in this case, come up with the notion of cutting the door into two halves.

3. Need for function and goal specifications

Some of the doors we came up with simply failed to function. That is, they failed to open and close. But how would an IDIS decide that a failure had even occurred? Well, we have an idea both of what a mechanism is supposed to accomplish (i.e. an overall goal) and how it is supposed to accomplish it (its function). In the case of doors, we already mentioned one functional specification based on our knowledge of how people use their arms to move objects. Since people only have two arms, we do not want to create a door which will take three or four arms to open. We also know that human arms are symmetrical. For instance, door A in figure 10 above might be appropriate for a race of aliens with asymetrical arm lengths (who only walk through doors in one direction!). The goal the door is to serve is also important. Is a door intended to be walked through or to serve simply as an access. For example, refrigerator doors are *not* intended to be opened from both directions. Furthermore, many refrigerator doors are cut asymmetrically, (vertically or horizontally) because the freezer part is smaller.

Thus, an IDIS needs a description of the functionality of the door. This serves as the basis for creative invention. This information is also essential in determining when to reject a door simply on the grounds that it is awkward.

That is, IDIS needs specifications or criteria of the sort:

> A door that can be opened by one simple motion or by two simlutaneous motions is better than a door which requires several serial motions.

For instance, bar doors can be opened by pushing with both hands simultaneously. However, our earlier combination trap door/bar door required too many motions to open.

The purpose of the door is as important as its function or use. Inventions often fail in terms of cost and market feasibility when compared against alternative solutions. We would want our IDIS to ask the questions:

> Is a door even needed here? Can the goal be solved by some method other than a door?

These questions require evaluating the overall goal that the designed object is intended to fulfill. For instance, if one needs access to an object only once, a door (which can opened and close repeatedly) may not be needed. A child's piggy bank, for instance, is made to be broken. The same question of a goal can be applied to each component making up a given design. Can we have a door without hinges, or without movement? Can we replace materials with other materials? This line of questioning, for instance, can lead to the design of doors such as those intended for pets, which are composed of overlapping pieces of triangular plastic, where the plastic bends back as the pet pushes against it. Here, hinges and rotation have been replaced by the pliability of the plastic. In any case, top level goals serve to direct and control creative processes.

4. Recognizing what is really novel

During these protocols, it is clear to us when a rule simply creates a variation and when it resultes in a truly new kind of door. For instance, merely changing the color, shape or handle placement of a door does not seem to create a novel door. However, cutting the door in half gave rise to two novel doors: accordian and bar doors. How do we recognize we have come up with something really novel?

An IDIS will need heuristics for recognizing novelty. Here are some:

> A novel design may have occured when:
>
> a) the way the new object is used is different
> b) the behavior of the object is altered, or
> c) the invention itself creates a new feature which can be combined
> with others to greatly increase the combinatorial possibilities.

For instance, altering door shape did not alter the way the door was opened, nor did it change the way the door was used, nor did it open up any new combinatoric possibilities (i.e. shape was already known as a possible feature to vary). However, the bar door was novel since, whereas before the door opened by one rotational arc, now the door opened with two rotational arcs, each made in the opposite direction.

The accordian door was a more dramatic invention, since the way it opens and closes is not rotational at all, but consists of a straight line motion (in the limit). In each case, a truly novel door results in a different use, or it opens up new possibilities for combination.

3.4. Episodic Memory and Analogical Reasoning

Notice that when we encountered the 3-slab bar door and noticed that the center slab was unattached, we immediately said "Let's try the solution we just used when inventing the accordian door". As a result, we got the accordian bar door. Here, recognition of a problem caused us to not only apply general problem-solving knowledge, but also to look at past solutions. This is a way of automatically increasing the problem solving power of an IDIS.

In fact, one thing we noticed was that an attempt to invent a novel door by use of c-rules often drew us immediately to recall a door that we, as humans who already possess alot of experience with doors, already were familiar with. For instance, one aspect of a door is the way in which it opens. For example, we assumed that our IDIS had the following heuristic:

C-rule 3.4a: (Function Alteration)
To create a new object O' from O, select a function of a component of O and alter it, using whatever methods are known.

Using this c-rule, we decided to select motion to alter. Instead of rotation, we selected forward motion and chose wheels as a known method. This approach resulted in the "wheel door" of figure 11.

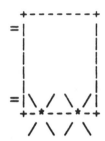

Figure 11: Door on wheels

One problem with this door was that it could not move. So we decided to remove the hinges and let the door roll in front of the its opening. One thing we noticed, however, was that the minute we thought of rolling a door sideways, we immediately recalled sliding-glass doors. Of course, an IDIS that does not already know about sliding doors will not be able to recall them, and must invent them from scratch. So we must still be concerned with how such an invention might take place in a system without a memory of prior inventions. However, the recall of known *relevant* inventions while one is trying to be creative is clearly a very common occurrence.

Up until now, we have assumed that an IDIS will possess semantic information about naive physics, mechnical operations, etc. However, the memories of human inventors are also filled with numerous design objects, both those discovered by the designer and those invented by others.

Whenever an object is designed, or whenever the inventor comes upon an objected designed by another, that experience is indexed and stored away in the inventor's episodic memory. As new objects are invented or encountered, they are also indexed in episodic memory. If a new object shares relevant features with an object in memory, then the inventor will recall that object.

In our case, rolling a door sideways on wheels shared enough relevant features with sliding doors that we immediately were reminded of that class of doors. This reminding process has both advantages and disadvantages for someone who is trying to be creative. Recalling prior designs has the advantage that these designs immediately become available for use to the inventor who is trying to solve a problem. It also has the advantage that the inventor need not constantly reinvent the same object over and over again, but instead can simply use recall in all subsequent cases after having first invented or encountered the object.

3.4.1. The Paradox of Creativity

The "paradox of the expert" in AI traditionally has been that the performance of computer expert systems often worsens as more rules are added to the system. In contrast, human experts usually get better at problem solving the more that they know. The cause of poorer performance in AI systems is that new knowledge is not properly indexed or organized into memory, so as a result it interferes with existing knowledge.

In the area of creativity, "the paradox of the expert" is something that human experts often *do* suffer from. That is, human experts often lose creativity of invention as they become more experienced in their own area of design. Computer systems have lacked this problem, not because they do not have it, but simply that there are no computer systems really capable yet of human creative thought.

With people, it is often the case that a new breakthrough will be discovered by a naive scientist or designer who is relatively new to the field. This new designer usually has been in the field long enough to know the techniques and important issues that drive the field. Beyond this requirement, however, often the novice designer comes up with the creative ideas while the established experts do not.

Why is new blood so effective? The answer has to do with the positive and negative consequences of experience and the way those experiences are organized in memory.

The expert knows hundreds of door designs already; thus saving him from having to reinvent them. However, it also becomes difficult to come up with a completely new type of door, since every time the expert applies a rule of creativity (such as altering a feature) it results immediately in accessing an already existing door or accessing an already existing unsolved constraint (i.e. one that had been abandoned). Now this is very efficient, in the sense that old work does not have to be redone. However, through time, circumstances change, e.g. new machines, new materials, new production techniques (often invented by others) alter the constraints on old, blocked discovery paths. This means that exploring these old paths can become effective. The new inventor starts this exploration without constantly finding in memory the old constraints that block access. Thus, while the new inventor has to do some reinventing, often a breakthrough is accomplished while the old inventors have given up saying "everyone knows that can't be done". There is a saying, even, that students will solve problems professors have not solved because the students are too ignorant yet to know that it cannot be done.

The second reason the novice inventor is effective is because he brings with him a slightly different episodic memory of experiences often from a distinct domain, and applies this experience to the new domain.

3.4.2. The Role of Analogy

One of the most powerful rules of creativity involves the use of analogies:

> C-rule 3.4a: (Analogical Mapping)
> To invent a new object O' given an object O in task domain D, recall an object O1 for task domain D1 in episodic memory which shares features with O, then modify O1 analogically to function in domain D.

The application of C-rule 3.4a resulted in reinventing a number of doors, including the revolving door. While thinking of other objects that "let things pass by, over, or through them" we suddenly were reminded of the water wheel. By turning the water wheel on its side, we arrived at the revolving door, where one person enters the door as another leaves it (figure 12).

Figure 12: Revolving door
(view from top)

Recalling objects from other domains, and then using these objects in the task domain, is a powerful stratgey powerful for invention. By looking at how other, non-door objects are handled, moved, swung, hinged, made stiff or pliable, etc. we get inspiration for new kinds of doors.

The most complicated process of invention we went through, and that lead to our most unusual door, resulted as follows:

1. We thought of an object that lets only part of something through it. This lead to being reminded of strainer which lets liquids pass through but capture solid objects of a certain size (figure 13).

Figure 13: Strainer

2. We attempted to alter it so that it would let objects of human size through. We tried the heuristic rule of eliminating some component of the object. We removed cross fibers. This resulted in a "rope door" -- i.e. a door of ropes attached at the top and bottom (figure 14).

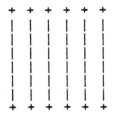

Figure 14: Elastic Rope Door

One problem with this door is that it cannot be 'closed' to block access. Attempts to correct this resulted in attaching wires to the rope so that the strands could become magnetized. As a result, the ropes would stiffen and not let larger objects pass through. This "magnetic sieve door" is a rather bizarre door. At this point its marketability is not the point. What is interesting is that thinking up such a door relied heavily on both a) recalling an object in an unrelated domain yet which shared certain critical abstract features with door and b) mapping this object into the task domain and then deforming it to satisfy constraints within the task domain. An IDIS capable of such processes would begin to approach human inventiveness and design skills.

We have seen that a strong component in creativity is the ability to take an object from a completely different domain and apply it to the current task domain, altering it as needed. This adaption ability presumes the prior ability of *recalling* or being *reminded* of this object in the first place. But what allows us to recall object B from domain D2 when we are thinking about object A in domain D1? Clearly, objects must be represented and indexed at varying levels of abstraction.

These levels must be abstract enough to allow for cross-contextual remindings, otherwise inspiration would never occur. This cross-contextual reminding process greatly increases the potential number of combinations available for analysis. However, if every object in memory were compared with every other object, then the combinatoric results would become overly undirected and explosive. We do not want to think of bicycles when we are trying to come up with a better tent. However, we *may* want to think of umbrellas. So our memory must be organized to allow cross-contextual remindings where some semblance of functionality or other features are preserved. This means that objects must be indexed and represented along meaningful lines. The notion of "letting things through" is very abstract, and yet allows for both sieves and doors to share indexing at this level. As a result, a sieve can be recalled and potentially used in the creative process.

4. Prerequisities for Design Creativity

We have seen that the process of inventive design for objects as simple as doors is rather complex from a cognitive modeling point of view. Key issues in design invention include: combinatoric search, episodic memory organization, cross-contextual remindings and analogical reasoning, a naive physics of the world, and problem solving using contraints. It turns out that these issues are *generic* issues in all areas of AI research. In a way, this fact should not be surprising. The task of AI is to model intelligent processes, and every human being is an   inventor and problem solver at some level.

What other capabilities will an IDIS need? Here are a few:

- The ability to acquire naive physical knowledge automatically through interaction and direct experimentation with the physical world.

- A capacity for visual imagination and manipulation -- i.e. knowing how objects move, twist, grate, tilt, bump, bend, slide etc. as various forces are applied. This imagination is dependent on internal representations that we can manipulate, even to the extent of imagining not only things which we have never encountered, but also in imagining impossible objects or objects which violate known physical laws (as in cartoons). Furthermore, we can combine images at will. It is easy to imagine an elephant, floating 10 feet off the ground, purple in color, upside down, with a dog walking underneath the elephant, backwards, wearing ski boots. A major research task is that of developing representations for the visual and physical proeprties of objects so that creative imagination is possible.

- Language acquisition ability and skill in comprehension of natural language (Dyer 1983) (Schank 1982). Most problems are communicated in a combination of language and technical notation, but the language component contains the major generic concepts concerning what is at issue. For instance, in computer science no one can read a computer program if the variable names are not mnemonic (i.e. refering to generic concepts accessible through natural language). Furthermore, people always read the documentation first, because it explains what the program "is about". The case is similar with schematic diagrams. Information conveyed in English which accompanies schematics often supplies the necessary conceptual context for understanding the diagrams.

- The ability to learn new rules of invention and creativity as knowledge of the domain grows. The ultimate IDIS must be able to go beyond naive physics to sophisticated engineering knowledge as more experience is gained.

5. Conclusions

The future leaves open the possibility that automated computer inventors might someday surpass the capabilities of human inventors. The automated inventor could be programmed to apply its rules of invention along previously explored paths every so often, and thus reexamine constraints in the light of new knowledge. Working in an exhaustive manner, hour after hour, an IDIS might be able to perform the combinatoric search, analogical mappings, constraint analysis and problem solving that many human inventors find so exhausting. Solving these probems in the general case, however, would be equivalent to having solved the complete problem of thought, language, learning and intelligence.

Recent work on machine learning and invention in Artificial Intelligence (Michalski et al. 1983) have shown that creativity and design are not impenetrable barriers. Many aspects of imagination, design and creativity *can* be formalized and encoded into a machine. This potential is very promsiing and exciting. However, the research work on creativity, discovery, and invention in Artificial Intelligence is only a decade old and should not be judged prematurely.

6. Bibliography

Dyer, Michael G. *In-Depth Understanding.* Cambridge, MA: MIT Press, 1983. (Artificial Intelligence Series).

Forbus, K. D. Qaulitative Reasoning about Physical Processes. *Proceedings of the Seventh International Joint Conference on Artificial Intelligence.* IJCAI-81,Vancouver, B.C. pp. 326-330, 1981.

Hayes, P. J. The Naive Physics Manifesto. *Expert Systems in the Micro-Electronics Age,* edited by D. Michie, Edingurgh University Press, Scotland, 1979.

deKleer, J. and J. S. Brown. Mental Models of Physical Mechanisms and Their Acquisition. *Cognitive Skills and Their Acquisition.* edited by J. R. Anderson, Erlbaum, 1981.

Langley, P., Bradshaw, G. L. and Simon, H. A. BACON.5: The Discovery of Conservation Laws. *Proceedings of the Seventh International Joint Conference on Artificial Intelligence,* IJCAI-81, Vancouver B.C. pp. 121-126, 1981.

Lenat, D. B. Automated Theory Formation in mathematics. *Proceedings of the Fifth International Joint Conference on Artificial Intelligence.* IJCAI-77, Cambridge, MA, pp. 833-842, August 1977.

Lenat, D. B. The Role of Heuristics in Learning by Discovery: Three Case Studies. *Machine Learning.* Tioga Press, Palo alto, CA, . 1983.

Michalski, R. S., Carbonell, J. G. and T. M. Mitchell (eds.) *Machine Learning: An Artificial Intelligence Approach.* Tioga Publishing, Palo Alto, CA, 1983.

Nilsson, N. *Principles of Artificial Intelligence.* Palo Alto, CA: Tioga. 1980.

Pearl, J. *Heuristics: Intelligent Search Strategies for Computer Problem Solving.* Addison-Wesley, Reading, MA, 1984.

Schank, R. *Dynamic Memory: A Theory of Learning in Computers and People.* Cambridge University Press, 1982.

CHAPTER 3

COMPUTER AIDED
PROCESS PLANNING

Reprinted from *Computer-Aided Engineering Journal*, December 1984

Artificial intelligence for production planning

by Frank Mill and Stuart Spraggett

Coventry (Lanchester) Polytechnic

Artificial intelligence tools have recently become a significant topic in many areas of engineering research. The use of computer-based decision support systems is potentially of great benefit in the complex task of process planning, both in terms of consistency of approach and in the reduction of manufacturing lead times. This paper discusses a problem-solving system which carries out process planning tasks for a small flexible manufacturing system.

Introduction

In recent years a great amount of effort has been spent attempting to automate various activities in the manufacture of small batches of mechanical components. Computer-aided design technologies such as automated drafting, solids modelling and finite-element analysis are making a major impact on the design departments of many firms, while computer-aided manufacturing advances in computer numerical control, direct numerical control, robotics, intelligent conveying, automatic inspection and the integration of these into complex control structures are changing the way small batches of components are made.

Between the design and make activities, however, lies the task of production planning which involves the interpretation of designs and the subsequent production of a set of manufacturing plans. Part of this problem involves efficiently scheduling the flow of work through the manufacturing system. It is in this area that substantial improvements have been made in recent times; these are a result of both mathematical advances and the existence of powerful computational and simulation tools. The process planning

activity, on the other hand, has gained much less from the availability of new technologies. The production of time and cost estimates for components is often aided by the use of computers, but attempts to produce systems which can make decisions about machine selection or the sequencing of cutting operations have resulted in only partial success, and these are usually specific to given application areas.

At the present time there is considerable interest in the application of artificial intelligence (AI) techniques in engineering. These techniques may be employed to produce diagnostic facilities on machine tools during manufacture, for example, while Simmons [1] has pointed out the possible benefits which might be realised if AI can be applied in computer-aided engineering design. In production planning too AI technology appears to be capable of making major improvements on existing systems.

The process planner's job involves reasoning about and interpreting engineering drawings. He or she has to make decisions on how cuts should be made and in what order the cuts should be executed, as well as deciding which ma-

chines and tools should be used and what fixturing is necessary for holding the component during cutting.

These decisions are often made as a result of subjective judgments on the part of the process planner involved and are prone to inconsistency. Not only are plans for identical parts likely to vary between different process planners, but they will also vary over time in the case of a single process planner. The type of plans which a planner produces depend on the individual's technical ability, the nature of his or her past experience and even on the person's mood at the time of planning.

Computers may be used to perform what would otherwise be time-consuming calculations, but they have had little impact on the decision-making functions which are involved in process planning work. It may also be argued that the decision-making process itself has become more complex for many planners, given the fact that manufacturing technologies change constantly and new options in tools and materials are continually being made available.

The process planner must possess a high degree of manufacturing knowledge to accomplish his or her task in a satisfactory manner. Thus the planner is often expected to have considerable practical experience as well as a high standard of formal education. Such people are likely to be difficult and expensive to recruit and hold for many firms.

Another problem which is often associated with process planning is the significant contribution it makes to manufacturing lead times owing to the time it takes to manually develop plans. The use of computer-based decision support systems in process planning could greatly reduce many of the problems discussed, while the technology needed to produce such systems may now be emerging.

Computer-aided process planning

The idea of using computers to help produce process plans was first discussed by Niebel [2] in 1965. Since then many efforts have been made to develop process planning systems and these have fallen into two distinct types, known as variant and generative systems.

The variant approach involves storing and retrieving standard sets of plans for parts which are usually classified into family groupings on the basis of their geometric shape. These systems are useful in situations which allow for a neat and simple coding of all the parts which are handled. The data management facilities made possible by the computer can greatly reduce the time and cost involved in producing manufacturing plans, as well as helping to ensure that the plans will be produced in a more consistent way.

As Steudel [3] has pointed out, however, the standard plans must be coded by an experienced process planner and will probably be used and maintained by him or her. Not only is this costly but the computer is only a tool in a manual process planning activity.

The generative method of process planning, on the other hand, involves the automatic generation of a unique plan for each component and so does not require the storage of standard routines. Instead tentative plans are generated and tested for suitability, and then the best alternative is chosen according to the system's optimisation criteria. Generative systems are usually dependent on the use of a very detailed geometric coding system for components, and as a result use more resources at the coding stage than variant systems.

Recent years have seen the development of several generative systems which have been created for specific application domains. Trusky [4] describes a system for a gear cutting firm, while Wysk [5] has developed the APPAS system for use in milling and hole creation applications.

Halevi [6] outlines a system which was designed to produce plans for cylindrical components and also points out the size of the sequencing and machine selection problem. For example, consider a component consisting of N independent and different features such as holes. These may be cut in any order and each may be produced on any of M machines. The total number of different possibilities P for manufacturing the component may then be given by:

$$P = N! \times M^N \qquad (1)$$

Thus, for a simple part with ten holes which is to be made in a workshop with ten candidate hole creating machines, there are 3.6×10^{16} alternatives. Even if each alternative could be evaluated in one millionth of a second it would take approximately 1000 years to compute all the alternatives so that the best plan could be picked out. In practice, however, it is possible to consider only a tiny fraction of the combinations by cutting out seemingly fruitless possibilities using simple rules of thumb. In the past, when researchers have developed generative process planning systems they have tried to write programs which find 'optimum' plans, although this approach has several drawbacks.

The calculations involved in the evaluation of cutting processes are often misleading because these are subject to substantial errors caused by large tolerances on the variables used in the analytical equations which are typical of process planning. For example, the equations or tables which are used to calculate the metal cutting conditions required for a suitable surface finish often use variables which have tolerances of plus or minus 50%.

But even if the production of an optimal process plan could be guaranteed there are still questions about the desirability of a single rigid optimal plan. Such a plan may be costly to find in terms of time and might not fit in well with the subsequent production schedule. Optimal plans may contribute little to a firm's profitability if a component has to wait hours or even days to get onto the machines which are required to carry out its optimum cutting sequence.

The requirements which may be placed on future process planning systems may be different from those of the past as computer-aided manufacturing technologies make for advanced production methods such as those embodied in a flexible manufacturing system (FMS).

One of the major benefits which can be realised by a successful FMS installation is the avoidance of excess work in progress and the correspondingly fast throughput times. In the manufacture of small batches of components this can only be gained if the work can be scheduled to flow efficiently through the manufacturing system. Thus the demands placed on the process planning activity by an FMS may have different priorities than those previously associated with process planning.

A process planner for an FMS should ideally take on a much more integrated role than might be expected in traditional manufacturing systems. Such a planner would operate on a more systemic concept of efficiency and might allow the manufacturing system to operate closer to its optimum while still producing efficient cutting plans (although these might be semi-optimal rather than true optimal plans). This would require that the process planner actively contribute to system efficiency by including in its decision-making logic some consideration of parameters used to measure the status of the manufacturing system (for example machine or tool availability). Alternatively, the planner might make a passive contribution by outputting alternatives to its first choice plan. This may allow more flexibility in planning the schedule for a production period or may allow a real-time scheduling system to choose alternatives in the result of breakdowns.

Artificial intelligence in production planning

A possible method of improving existing process planning and scheduling activities lies in the technique of enhancing present methods with the addition of intelligent knowledge-based systems. Considerable interest in the use of these and other artificial intelligence tools has recently come about, and currently much research is being geared towards the use of intelligent knowledge-based and expert systems in engineering applications.

One job shop scheduling system has been described by Fox et al. [7], while articles discussing the possibilities of the use of artificial intelligence techniques in process planning work have recently appeared. Nau and Chang [8] have discussed the application of expert sys-

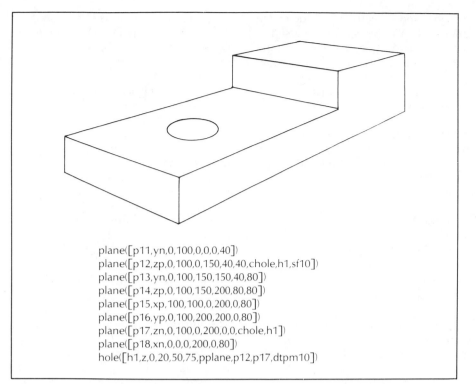

```
plane([p11,yn,0,100,0,0,0,40])
plane([p12,zp,0,100,0,150,40,40,chole,h1,sf10])
plane([p13,yn,0,100,150,150,40,80])
plane([p14,zp,0,100,150,200,80,80])
plane([p15,xp,100,100,0,200,0,80])
plane([p16,yp,0,100,200,200,0,80])
plane([p17,zn,0,100,0,200,0,0,chole,h1])
plane([p18,xn,0,0,0,200,0,80])
hole([h1,z,0,20,50,75,pplane,p12,p17,dtpm10])
```

Fig. 1   Component and its part description

tems techniques using hole creation as an example. Hannam and Plummer [9] describe the successful work carried out by them in attempting to build 'good production engineering practice' into a process system for turned parts.

An expert system for process planning called GARi is the result of research described by Descotte and Latombe [10]. This system consists of a planner and a 'knowledge box'. The 'box' contains the manufacturing rules which are used to guide the search for a solution to the process planning problem for a given component.

Research carried out by the authors has resulted in a problem-solving system which carries out process planning tasks for a small flexible manufacturing system. Like other artificial intelligence problem solvers the model is made of three components: a database, a set of production rules, and a control strategy. The database holds details of part geometry, blank-material geometry and the machines available, as well as storing details of the cuts to be taken.

The production rules consist of a series of operators which perform transformations on the database in order to achieve specified goals. Such rules are statements of knowledge about the problem domain and how parts of a problem may be solved. The decisions on which of these rules to apply and in which order are a matter for the procedural semantics of the system and

it is the operation of these which implements the control strategy of a program.

The model starts by using the part description as a definition of a problem and regards the description of the blank material as a goal state. Thus the system attempts to transform the part description into the blank description while the details output by the transformation are a series of metal additions to the component. These are used to represent a reversed and tentatively sequenced list of cutting operations. This list is then put into a final suggested sequence of operations and the best machines to carry out the operations are also assigned, on the basis of a complex cost system. Alternative plans and options also result from the planner.

In the model, both the part to be produced and the blank material from which it is to be cut are represented by the same descriptive technique. This allows any shape of blank material to be specified to the system. Such blanks may include partially completed components which are a result of a casting or other process for example.

The description technique used is based on the general geometry of the component and infers little about the manufacture of the component. In using the technique the part is described as a list of features, such as pockets, slots, planes or holes for example. The description of a feature itself

consists of a list of attributes as shown in Fig. 1. The list contains such information as feature name, parent features' names, child features' names, tolerances, the names of features to which these tolerances refer and a geometrical description of the feature itself.

The description only gives rough geometrical information; in the case of a plane, for example, only the minimum and maximum distances from the axes are given. This is sufficient for planning purposes since details of the exact shape are not necessary until part programming is being undertaken, while any difficult areas requiring special attention are referred to as separate child features.

The description of machines in the system is done in a similar way to the part description, whereby the machine is represented by a capability profile which gives details of the shape creation facilities available. As in part description a list is included detailing maximum sizes, tolerances achievable and the cost of using the machine.

The planning model is composed of a number of modules, each of which has its own set of heuristic information. This is stored in the form of simple rules about how to perform a given task under set conditions. These rules may be regarded as being in the format of IF-THEN type statements. For example, the first step the model takes is to give a tentative order to the features on a component. This is done purely by applying a rule to the component description. These rules may be expressed in a verbal form:

IF the feature being considered is referenced to in tolerances
THEN accept the feature onto the new feature list
ELSE consider another feature.

After the execution of such a rule another may be applied to order the remaining features. It is these rules which collectively make up the knowledge bases in the model.

Although the rules described above make up the knowledge of the model, their efficient implementation depends on the control strategy of the program. The way in which rules are applied depends on the order in which they are placed and on the operation of special control rules which transfer control from one part of a program to another. It is this control technique which implements the search strategy for a solution to a problem. Many different types of search strategy have been developed in order to search efficiently for solutions to problems and these vary enormously in both their scope of selection (the extent to which alternatives are considered) and their scope of recovery (the extent to which they can reverse previously made decisions).

In attempting to solve a number of complex problems such as those encountered in the process planning activity, it is unreasonable to try to use a single solution search strategy, and so a number of different types may be combined depending on the subproblem at hand and a unique hybrid strategy may be formed.

This is the approach which has been adopted in the building of the process planning model, and indeed it is the ability to experiment with different strategies which is the reason for the model's development.

After initially sorting the features of a part into a tentative order, the model can then choose a selection of suitable machines for each cut to be taken. The initial choice for a given feature is done in the following way. The model checks the parameters given in the machine capability profile and discards machines which cannot perform the given task; then the choice is made on a cost basis, the cheapest M machines being suggested and filed with the cut details. If only the cheapest three machines are considered, this has the effect of setting $M = 3$ in Eqn. 1, giving:

$$P = N! \times 3^N \qquad (2)$$

Thus the search space has been reduced by removing the need to search seemingly expensive possibilities.

When this selection procedure is complete it is then possible to make decisions on the sequencing of operations and on the actual machine to be used for each operation. The search space for this subproblem may be immense, and a great deal of computation may still be required if an optimum solution to this subproblem is to be found, and at this stage heuristic information is used to narrow down the possibilities. If, for example, a single anteriority rule can be applied such that a certain cut A must be performed before a cut B, then this could reduce the number of combinations in Eqn. 2 to:

$$P = (N!/2) \times 3^N$$

Any further anteriority rules which can be applied would further reduce the size of the search space.

Rules are subsequently used to group operations into set-ups and a simple al-

though crude method of detecting groups is applied. This uses the assumption that any cuts being accessed in the same direction may be performed in the same set-up. Next, the model finds subgroups within set-up groups, which are operations which can be done on the same machine and in the same setup. The model can now detect the number of machine-to-machine movements and set-ups required and assigns costs for these in the plan.

The final step in this stage of selection consists of considering alternatives. From the list of alternative second- and third-choice machines the model tries to substitute machines into whole phases of operations and tests to see if the increase in machine cost can be more than compensated for by a possible decrease in the number of machine-to-machine movements required. Following this the choice of machines is set, as is the sequence of phases.

Final tool selection is not carried out by the model for two reasons. First, it was considered that the task of tool selection may be done in a way similar to that of machine selection, and secondly the argument that tool selection should be done as late as possible in an FMS is accepted. This allows decision making about tools to be carried out with regard to the production schedule and the tool sets to be used. Calculations on speeds and feeds of the cutting process can be done after tool selection, possibly using a computer-based machineability data system.

Costs

As mentioned earlier, manual methods of process planning often make use of the planner's general knowledge of the current and past production schedules, and possibly even of future ones. Components can be directed to machines which are not usually too busy, thus easing or smoothing the production schedule. In the event of things going wrong the workshop foreman may also change machines or alter plans. In the model described, machines and sequencing were selected on cost criteria. The costs used in the model are not set in a conventional manner, however. In addition to machine costs per hour, additional weighted costs are added or subtracted from a feedback in the production schedule.

This might be used in a number of ways. It is possible to adjust the cost of a machine depending on its state of loading. When a nominal production

schedule is set it is possible, also, to rerun a process plan for one or more components and make an overloaded machine more expensive so as to reduce the process planner's propensity to use that machine. If one wished to exclude a machine owing to breakdown, for example, then that machine's cost can be set to infinity, thus effectively excluding it from process planning consideration.

The model also generates alternatives for each operation and phase if these are possible, and this might allow a scheduler some flexibility. In real-time control of the manufacturing facilities the use of alternative machines may allow a component to be redirected to other machines in the event of breakdowns.

The future

The model which is described in this paper was originally intended as an experiment to see if artificial intelligence techniques might prove useful in automating some process planning tasks and also to help the researchers develop an understanding of what heuristic techniques might be used. To this end the model has been very successful and it is intended to develop the model more fully. There is much work which could be done in this area.

While the model was being developed the part description format evolved as a result. Some work has now been done on considering a formal approach to coding parts using a hierarchy of features. However, great benefit might be gained if codes for process planning could be generated automatically by an interpreter interfaced to an engineering database.

The model could also be improved by providing a larger number of features, as well as significantly more complex ones such as planes which do not lie parallel to any axis, or possibly sculptured surfaces.

The model uses a simplistic method for reasoning about set-ups, and it would be useful to investigate the design of a more realistic and correspondingly more sophisticated technique. Although outside the scope of the present work, it would be worthwhile to investigate the possibilities of fixture selection and positioning using artificial intelligence techniques.

It will take some time before the process planning activity can be automated; however, the techniques which might be used to achieve this goal may now be beginning to emerge.

Implementation

The model which has been described in this article was written in C-Prolog running under the UNIX† operating system. The scheduling work is manually prepared and is verified on a simulation of the FMS on an Istel SEE WHY system. Tool paths are generated and tested on a Micro Aided Engineering NC machine tool simulation system.

Acknowledgments

The authors would like to give sincere thanks to the following persons: at Coventry (Lanchester) Polytechnic, Jim Davis of the Department of Production Engineering, and Alan Chantler, Nick Godwin, Simon Ritchie and Rob Lucas of the Department of Computer Science; and Nigel Kay of the National Engineering Laboratory.

† UNIX is a trademark of Bell Laboratories.

References

1 SIMMONS, M. K.: 'Artificial intelligence for engineering design', *Computer-Aided Engineering Journal*, 1984, **1**, (3), pp. 75–83
2 NIEBEL, B. W.: 'Mechanised process selection for planning new designs'. American Society of Mechanical Engineers, 1965, Paper 737
3 STEUDEL, H. J.: 'Computer-aided process planning: past, present and future', *International Journal of Production Research*, 1984, **22**, pp. 253–266
4 TRUSKY, H. T.: 'Automated planning reduces costs', *American Machinist*, 1984, April, pp. 80–82
5 WYSK, R. A.: 'An automated process planning and selection program: APPAS'. Ph.D. Thesis, Purdue University, West Lafayette, IN, USA, 1977
6 HALEVI, G.: 'The role of computers in manufacturing processes' (J. Wiley, 1980)
7 FOX, M. S., ALLEN, B. P., SMITH, S. F., and STROHM, G. A.: 'ISIS — A constraint-directed reasoning approach to job shop scheduling: system summary'. Report CMU-RI-TR-83-8, Carnegie-Mellon University, Pittsburgh, PA, USA, 1983
8 NAU, D. S., and CHANG, T. C.: 'Prospects for process selection using artificial intelligence', *Computers in Industry*, 1983, **4**, pp. 253–263
9 HANNAM, R. G., and PLUMMER, J. C. S.: 'Capturing production engineering practice within a CADCAM system', *International Journal of Production Research*, 1984, **22**, pp. 267–280
10 DESCOTTE, Y., and LATOMBE, J. C.: 'GARi: a problem solver that plans how to machine mechanical parts'. Seventh International Joint Convention on Artificial Intelligence, Vancouver, Canada, Aug. 1981

F. Mill and Dr. S. Spraggett are with the Department of Production Engineering, Coventry (Lanchester) Polytechnic, Priory Street, Coventry, Warks. CV1 5FB, England

Presented at the CASA/SME AUTOFACT 6 Conference, October 1984

A Knowledge-Based System for Machining Operation Planning

by Brian E. Barkocy
and
William J. Zdeblick
Metcut Research Associates Incorporated

INTRODUCTION

Knowledge-based systems, which capture human problem-solving expertise in computerized form, are being developed for many planning and decision-making functions within automated manufacturing, including facilities design[1], process planning[2,3,4], robot control and maintenance[5]. These "expert systems" provide opportunities for productivity improvement by making valuable knowledge available to a wide range of users, who otherwise might have little access to the cumulative experience and proven techniques in their field. One application area, however, which has not been widely addressed by knowledge engineers in manufacturing, is operation planning for machining. Operation planning is defined as the specification of parameters for the individual operations within a part's process plan[6]. For machining operations, this activity includes selection of the cutting tool, cut sequence, cutting conditions, tool replacement strategy, etc., to produce a single feature (slot, hole, etc.) on a part. In contrast, process planning is a broader activity involving selection of operations and processes needed to produce multiple features on a part and transform the raw material into a finished part. While important work is being done in applying artificial intelligence techniques to process planning[2,3,4], the economic importance of operation planning also calls for the application of knowledge engineering. It is estimated that the labor and overhead cost for machining required for manufacturing in the USA is of the order of $125 billion annually[7]. The potential cost savings of improved detail planning of machining operations is, therefore, very large.

This paper presents a knowledge-based system, called CUTTECH, which captures metalcutting technology and data for use in recommending productive and economical tools and cutting parameters for machining. The system's functions include selecting cutting tools, cut sequences, speeds and feeds for a user-defined part feature to be machined. The value of such a system is underscored by the fact that, as NC machining continues to become the more popular method for batch manufacturing, the ranks of experienced conventional machinists are declining through promotion and retirement. The loss of shop floor expertise mandates the development of computerized system to capture machining knowledge. As illustrated in Figure 1, the CUTTECH systems acts as both a knowledge source and a knowledge collection point in a machining facility. The center of the

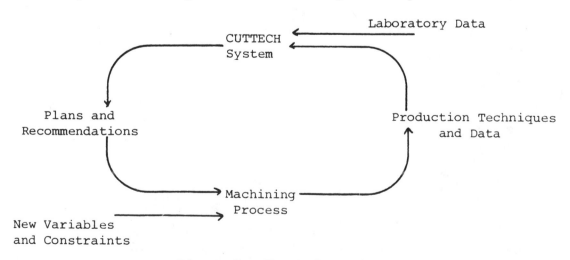

Figure 1: The Information Loop

"information loop" is the actual machining process, since it generates real production data as metalcutting takes place. Providing input to this process, in the form of an operation plan, is the CUTTECH system. The machining process in turn generates new techniques and data as plans are modified to meet actual production requirements. These new techniques and data are documented, and along with laboratory data become inputs to future production processes by their incorporation into the CUTTECH system.

The CUTTECH approach differs from most artificially intelligent planning systems in that it adds intelligence to a core of machinability data, rather than adding practicality to abstract geometrical concepts for machining (see Figure 2).

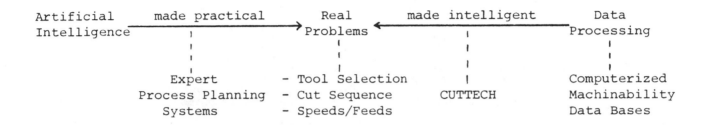

Figure 2: Approaches to Machining Planning

Systems which plan operation sequences for parts based on CAD data[2,3,8] address the problem of machining from a geometric perspective, while CUTTECH does so from a machinability viewpoint. CUTTECH builds upon basic machining data[9,10], increasing its value by using it as input for knowledge-based rules[11]. The effect is that the system provides "shop floor" intelligence - speed/feed and cutting tool recommendations - in terms both recognizable and maintainable by production-oriented personnel. This "bottom-up" approach to implementation of intelligence for machining allows for automation of repetitive, lower-level decisions, so that planners and NC programmers may address higher-level issues of machine tool selection, operation sequencing, etc. The approach allows control over the introduction of intelligent systems in the planning function, since high-level decision modules need not be implemented until low-level modules have been tested and accepted. Another feature of the system is its generative method of problem-solving. The system captures an unlimited mix of input conditions in group technology codes representing a part feature and work material, then invokes the machining rules based on certain portions of these codes. Because the resolution needed to access decision rules is finer than that needed to access "similar" plans, as in variant planning, CUTTECH gives recommendations based on machining principles rather than machining history[12]. The outputs generated by CUTTECH are not "canned solutions" to predefined problems, but a series of small solutions to each subproblem (tool selection, cut sequence, speed/feed selection, etc.). These solutions interact synergistically to provide the overall solution, i.e., the operation plan.

Following are an overview of the CUTTECH system and a description of each of its components, based on prototype systems built for several industrial companies. A sample set of terminal screens representing a user session is also included.

SYSTEM OVERVIEW

The CUTTECH system consists of an information gathering module, a knowledge base of machining rules, a controlling program, and a database of machinability and tooling information (See Figure 3).

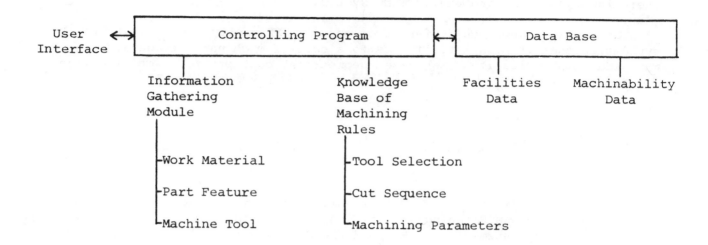

Figure 3:   System Components

The information gathering module consists of menu-driven decision trees which develop group technology codes to represent the workpiece material, part feature to be machined, and required machine tool capabilities for the operation to be planned.

The knowledge base of machining rules contains algorithms and decision tables for selecting the proper cutting tool, cut sequence, and machining parameters for the operation defined by the group technology codes. The controlling program governs the module sequence, user interface, and calls to the database which contains both machinability data (speed/feed recommendations, empirical constants, etc.) and facilities data (available machine tools, cutting tools, etc.).

A process planner or NC programmer begins a user session by describing a machining problem in the information gathering module, for example, end milling a slot in 4340 steel, on a part requiring a 3-axis machining center with 30 inches x-axis travel. CUTTECH's outputs include a list of the appropriate machining centers, a recommended end mill from the facility's tool catalog, and a table of feeds, speeds, and depths-of-cut for the necessary entry, roughing, and finishing passes to produce the slot. The CUTTECH system makes it possible for any user, even one with little machining experience, to obtain sound machining recommendations for a given part feature. The system's approach, that of capturing machining experience and data in an accessible, automated form, has been applied to milling, holemaking, and turning operations, and is especially suited to production environments characterized by:

- loss of experienced machining personnel

- dynamic part mix and frequent exposure to new machining situations

- need for a central focus for machining data and techniques generated in production or laboratory settings

Advantages of the component structure outlined above include procedural flow and modularity. Although the system is knowledge-based and performs "expert" tasks, it is designed for traditional procedural-language programming and fits easily into the factory data processing environment. Modular construction allows individual components to be improved or added without affecting the performance of other modules. Also, different knowledge representation methods (e.g., decision tables, rules, algorithms, data files) can be used for different functions, so that each best emulates human problem-solving in the particular domain.

INFORMATION GATHERING MODULE

The first task of any knowledge-based system is to learn what problem the user is trying to solve. CUTTECH accomplishes this through a series of inter-active decision trees in which the user answers questions in standard machining terminology to describe a workpiece material, part feature, and machine tool requirements for an operation. For example, to describe a work material, the user chooses from a list of materials (stainless steel, alloy steel, aluminum, titanium, etc.), then from alloys of the chosen material (4340, 4140, etc.), then from available heat treatments/hardnesses, and so on until he identifies a unique material. By answering questions, the user fills in the values for each digit in the group technology code for work material. The user then describes the part feature, choosing from options such as type (hole, slot, pocket, etc.) and sub-type (through slot, blind slot, etc.), and entering numerical dimensions for width, depth, fillet radius, etc. These responses define digits in the part feature code, which not only provides a basis for selecting the proper type and size of cutting tool, but is easily linked to a part/operation database or process planning system. Thus, the user could link a specific CUTTECH session to a part and operation number, or even download a feature code created previously for that part in a process planning system.

Experienced users of CUTTECH can reduce material and part feature coding time by "teaching" the system key words which represent commonly-used series of digits. For example, a planner who frequently works with 316 stainless steel may teach CUTTECH (in a special maintenance routine) that the key word "316" corresponds to the material code for that alloy. Then, in subsequent CUTTECH sessions, the user "shortcuts" the material coding tree by responding "316" to the first question, whereupon CUTTECH retrieves the material code for 316 stainless steel and proceeds to the next function. Of course, the "tutorial" mode of answering each material question in sequence would still be available for novice users, making the system adjustable for different experience levels.

To select a machine tool, the user first answers questions to code machine characteristics required due to part size and geometry, such as number of axes, number of spindles, table size, etc. CUTTECH uses the code to retrieve machines of the proper characteristics from a data file of available machines. The user quickly identifies candidate machines and selects the proper one for the operation and CUTTECH stores the machine identifier along with the material and part feature codes. When possible, CUTTECH speeds up the machine selection process by using the material and part feature codes to predefine machine characteristics and reduce the number of questions to the user. For example, if the part feature is a pocket and the available machine tool types are machining centers and vertical turret lathes, CUTTECH automatically defines the machine type as "Machining Center." If the operation were drilling, however, which can be performed on either type of machine, CUTTECH would ask the user to choose machine type.

KNOWLEDGE BASE

Once all information about the machining problem is acquired, CUTTECH accesses the knowledge base of machining rules to select the cutting tool, cut sequence, and machining parameters which constitute the operation plan. Rules are invoked based on combinations of digits present in the operation description codes. For example, if the material and part feature codes indicate pocketing in aluminum, then "plunge-cutting" and "high-speed steel" may become requirements for the cutting tool to be selected. The rules are stored in decision tables and algorithms, which can be modified to reflect different machining practices at different facilities (high-precision, high-volume, etc.). The contents of the rules may differ from plant to plant, but the generative structure is the same: a problem is addressed by solving its component sub-problems step-by-step. Rules accessed for each sub-problem are programmed to interact compatibly to form the overall solution.

Cutting Tool Selection

To plan an operation, the system first generates the necessary cutting tool characteristics, based on the chosen work material, part feature, and machine tool. These characteristics, such as minimum and maximum diameter, preferred tool material, etc., are used to weed out tools in the cutting tool data file which cannot or should not be used to perform the operation. (If CUTTECH has insufficient information to define a certain characteristic, it may ask the user to input acceptable values, or may even recommend a likely answer. For example, to define the style of tool for drilling a large diameter hole, the system may recommend spade drills instead of twist drills, but leave the choice to the user.)

The set of feasible tools having the necessary characteristics is then analyzed to find the most productive, economical tool which can perform the operation within the constraints of rigidity, surface finish, etc. The system uses rules and numerical heuristics to rank the tools by factors such as cost, machining time, metal removal rate, rigidity, etc. Since these factors interact in a complex fashion, the ranking rules are general in nature, reflecting the overall judgement of planners who frequently select tools under complex input conditions. Rules are applied in descending order of importance to sort a list of tools from most to least preferred. For example, "fewest passes required" may be the most important rule, with cutters scoring equally or nearly equally being evaluated further by a less important rule, such as "most rigid". The user may also assign weights of importance to the rules to compare heuristic "scores" of the ranked tools.

To complement the qualitative nature of these "rules of thumb," CUTTECH can also calculate estimates of quantitative factors, such as machining time and machining cost, for several tools in comparison. Although only estimates, the quantitative values provide additional insight to a user comparing, for example, a costly carbide tool versus an inexpensive HSS tool with a shorter tool life. The system can also provide such time and cost estimates for a tool pre-selected by the user, or for an "ideal" tool recommended by the system for comparison with the "best available." Thus, CUTTECH provides not only advice in the form of rules, but functional tools which enable the user to perform cost/performance tradeoffs himself when he has special requirements not addressed by the system. The system provides intelligence only up to the level required by the user, for example, selecting a specific tool for a novice planner while evaluating several tools for an experienced planner.

Cut Sequence Selection

After a cutting tool is selected, CUTTECH applies cut sequencing rules to divide the machining of a part feature into entry, roughing, and finishing passes as needed. The rules are based on a facility's standard machining practices, such as minimizing the total number of passes, making all rough passes of equal size, etc. Embedded in the rules are variables, called technology parameters, whose numerical values may be modified to change the rules' performance as new data and technology become available. Such parameters may include the maximum allowable ratio of cut depth to cutter diameter for end milling, or the recommended depth of pass for finish turning a certain alloy. These empirical values represent proper machining practices in a standardized and updatable form, and provide users with practical machining constraints to use along with geometrical constraints in planning a cutting tool path.

Machining Parameter Selection

The CUTTECH function closest to the factory floor (at the lowest level of decision-making) is selection of speeds and feeds for individual passes in an operation. Although much work has been done on methods to calculate economically optimal speeds and feeds[13,14], the production environment may not be suitable for analysis of each operation and work material due to time constraints or lack of access to such models by shop personnel. To provide speed and feed recommendations which are reasonable (although perhaps not optimal) and in a form understandable to production personnel, CUTTECH uses data tables containing guideline speeds and feeds for the various machining operations and alloys. These guideline parameters represent data known to be acceptable for the production environment. As improvements are made to such data through production feedback, the tables can be updated so that improved data is available to all users of the system.

To enhance the data tables' applicability to particular machining situations, the system applies rules which modify the guideline speeds and feeds when the operation description codes indicate severe machining conditions. For example, conditions such as excessive tool overhang or thin-wall part sections, which affect rigidity and deflection, could invoke percentage reductions in feed rate. Other conditions might invoke only printed warning messages or suggestions to the user such as, "Reduce feed in cornering pass due to severe cutting forces." As new data becomes available, verbal messages can become numerical modifications and empirical factors can be replaced by more sophisticated models. Thus the CUTTECH system can "grow" in intelligence according to user needs.

CONTROLLING PROGRAM

Tying the input and rule-based analysis modules of CUTTECH together is a controlling program which calls modules in proper sequence, communicates with the user and database, and coordinates knowledge flow through the system. The program stores operation description codes for access by the decision rules, and stores each individual step of the operation plan as it is determined. This enables the system to "recap" the machining situation when it presents its machining recommendations. To adjust the system functions and outputs to the user's level of machining expertise, the program provides the choice of "expert" or "novice" mode for functions such as cutting tool selection. In "expert" mode, switches in the rule access algorithms would be set to cause uncertain decisions to be presented to the user for approval or override, whereas "novice" mode would assume the system's decisions are likely to be more accurate than the user's. This enables the user to decide whether to use CUTTECH as an advisor and knowledge source, or as an assistant and data-manipulation tool.

Since certain CUTTECH functions provide valuable information outside the context of the complete operation planning system, users may access these functions independently. The controlling programs allows individual access to such functions as machining time and cost estimation, cutting tool data retrieval, and speed/feed data retrieval. A related feature is the ability to return to a previous system module during a planning session, to change a decision which has caused difficulty in a later module, or to modify choices slightly to obtain comparison outputs of several different machine tools or cutting tools.

Communication with other computerized systems, such as process planning, routing, tool inventory, and machine scheduling is also performed by the controlling program, as well as calls to the machine tool, cutting tool, and machining parameter databases. The emphasis of the program is to integrate knowledge from various sources so that each type of knowledge is used effectively. User inputs, decision rules, user override of rules, databases and other computerized systems are all called upon to create a viable operation plan. As new modules and functions are developed, they too can be integrated into the multiple-knowledge-source framework provided by the controlling program.

DATABASE

The CUTTECH system database contains two kinds of information needed for machining operations planning. The first is facilities data, representing the production tools available at a plant, such as machine tools, cutting tools, cutting fluids, etc. CUTTECH data files are computerized tooling catalogs containing detailed information (size, geometry, cost, etc.) about each tool available for use by manufacturing personnel in a plant. Updates to this data reflect changes to the plant's physical stock of tools, so that the data files contain only equipment currently available for production use.

The second type of data in CUTTECH is machinability data, a more subjective class of information representing cumulative knowledge of experienced machining personnel in a plant. Recommended speeds and feeds may be stored for any combination of operation, work material, tool material, and depth-of-cut. Technology parameters reflecting machining practices may also be stored and made available for easy review and updating. Sophisticated models may be built and referenced for tool life and tool breakage, and past CUTTECH planning sessions could be stored as a reference for planning similar operations or as a basis for comparing actual production data against original system recommendations when a feedback link is in place. The machinability database functions as both a source of recommendations and a collection point for production experience, and can document and capture a plant's valuable knowledge resources.

The features of both types of data in CUTTECH include:

1. Data are specific to one facility. A plant's own tools, machines, speeds, feeds, and machining practices are reflected, not simply industry standards or averages.

2. Data are modular, so that branch facilities can "plug in" their own tooling and machine data while using the same machining rules as other branches.

3. Data are maintainable, so that tooling and planning personnel can make new information and technology quickly available to all system users.

4. Data may come from files or data bases external to the system, through user-transparent interface routines.

SAMPLE USER SESSION

The following pages represent the user-computer dialogue during a CUTTECH session User entries are shown after the prompt (>) symbol, and questions are worded as they would appear on a terminal screen. Comments are included outside of the box representing each screen. The example shows the basic use of CUTTECH for planning a face-milled feature on a titanium part.

```
          Choose Type of Workpiece Material

          1  Alloy Steel
          2  Stainless Steel
          3  Aluminum
          4  Titanium

          > Ti334A

          Material Selected:  Spec. Ti334A   Ti-6Al-4V   Wrought 350-400 Bhn

          Choose Operation

          1  Milling
          2  Drilling
          3  Turning

          > 1
```

The user types in a "code word" for work material to bypass the step-by-step
coding process. He then identifies milling as the major operation type.

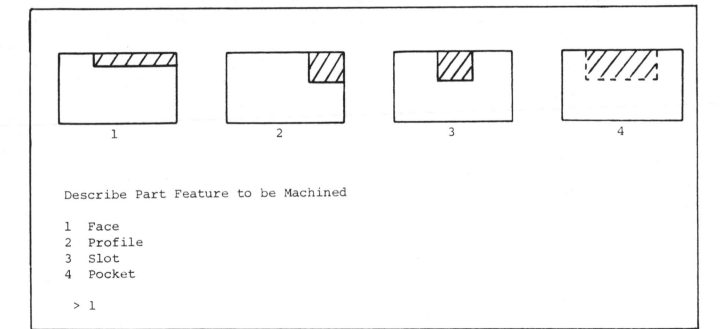

```
Describe Part Feature to be Machined

1  Face
2  Profile
3  Slot
4  Pocket

  > 1
```

Graphics aid the user in describing the specific part feature.

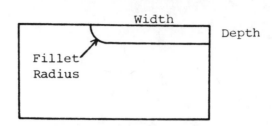

```
Enter Feature Dimensions (in inches)

Width, Depth, Length
> 3.5, .450, 10.0

Fillet Radius (min, max)
> .100, .150
```

The user inputs dimensions from the part blueprint, which become part of the operation description code.

```
Select Critical Machine Tool Requirements

1   Machine Type
2   Travel Limits
3   Table Size
4   Number of Axes

> 1, 4

Select Machine Type

1   Machining Center
2   Conventional Milling Machine
3   Gantry Mill

> 1

Input Number of Machining Axes Required

> 3
```

Machine tool requirements are gathered in order to search for acceptable machines.

```
Available Machines Are:

     ID              Make              Model          Overhead Rate    Shop Location
   ─────────────────────────────────────────────────────────────────────────────────
1  HURC001    Hurco                MBII Mach. Ctr.       38.00           L-10
2  MAZ01      Mazak                H-15 Mach. Ctr.       45.00           C-20
3  KT01       Kearney & Trecker    Data Mil 2430         35.00           D-09
4  OMNI02     Sundstrand           Omni-Mill OM4         40.00           D-12
5  SUN02      Sundstrand           OM-21                 38.00           F-10

Select the desired machine or enter 0 to search again

> 2
```

The user completes the information-gathering module by selecting a machine tool.
CUTTECH now applies decision rules to search for and rank cutting tools to
perform the operation described.

```
Available cutting tools are presented below.  Tool #1 is recommended for this
operation.
                                                                       Est.
                                                                     Cutting
Item    Catalog                        Flute      Tool     # of       Time
 #        ID        Type     Diam.     Length     Mat'l    Passes     (min)
─────────────────────────────────────────────────────────────────────────────
 1      FM-5579    Face Mill  4.000    0.400      C-2        3         6.3

 2      FM-4332    Face Mill  3.000    0.400      C-2        6         9.7

 3      EM-8732    End Mill   3.000    2.000      T15        4        37.8

 4      EM-9465    End Mill   1.250    1.250      T15        6        28.4

Select one for display of passes, speeds and feeds

> 1
```

Cutting tools are displayed in their order of applicability for the job as
determined by CUTTECH. Note the estimates of number of passes and cutting time,
which are normally tedious to calculate but helpful for comparing tool performance.

```
                         OPERATION PLAN


Material:  Spec. Ti334A   Ti-6Al-4V   Wrought 350-400 Bhn

Part Feature:  Face  Width = 3.500  Depth = 0.450  Length = 10.000
                     MinFR = 0.100  MaxFR = 0.150

Machine Tool:  Mazak  H-15  Mach. Ctr.   ID = MAZ01  SHOP = C-20  OVHD = 45.00

Cutting Tool:  ID = FM-5579   Diam = 4.00        FL = 0.400
                              Type = Face Mill   Mat'l = C-2
```

Pass #	Pass Width (in)	Pass Depth (in)	Cum. Width (in)	Cum. Depth (in)	Feed IPM	Feed IPT	Speed RPM	Speed SFM	MRR (cu in/ min)	Cutting Time (min)
			2 Roughing Passes							
1	3.500	0.200	3.500	0.200	5.52	.006	115	120	3.86	1.81
2	3.500	0.200	3.500	0.400	5.52	.006	115	120	3.86	1.81
			1 Finishing Pass							
3	3.500	0.050	3.500	0.450	3.68	.004	150	160	0.64	2.72

```
               Estimated Total Cutting Time:  6.34 min.
```

The final system output summarizes the operation description and presents the
recommended machining passes, speeds and feeds to produce the part feature.
The output parameters have been developed by applying cut sequencing rules and
machinability data.

SUMMARY

    This paper describes a knowledge-based approach to machining operation
planning and the benefits of its use. Although the approach involves conventional
programming methods, it contains to some extent the artifical intelligence
techniques of situation recognition (through group technology codes), heuristic
search (in finding cutting tools), and evaluation functions for possible solutions
(in ranking tools). The result is an implementable system which emulates human
planning decisions. The thrust of the CUTTECH system, as compared to other
artificially intelligent machining planning systems, is shown in Table 1.

Table 1:   Features of Planning Systems for Machining

AI Planning Systems	CUTTECH System
Practical framework for theoretical knowledge	Intelligent framework for practical knowledge (machining data)
Geometric Modelling	Group Technology Coding of Common Part Features
Parameter selection based on geometrical constraints	Parameter selection based on proper machining practices
Automate process planning decisions (made long before production occurs)	Automate operation planning decisions (made shortly before production occurs)
Emphasis on optimization	Emphasis on implementation
Theoretical in Nature	Practical in Nature

Clearly, all of the above features are needed for planning in the automated factory. The CUTTECH philosophy incorporates those features most closely related to present machining knowledge.

BIBLIOGRAPHY

1.   Fisher, Edward L., Shimon Y. Nof, "FADES: Knowledge-Based Facility Design". Proceedings International Industrial Engineering Conference, Chicago, Illinois, Institute of Industrial Engineers, 1984.

2.   Wolfe, Philip M., and Kung, Hsiang-Kuan, "Automating Process Planning Using Artificial Intelligence," Proceedings Annual International Industrial Engineering Conference, Chicago, Illinois, 1984.

3.   Matsushima, K., N. Okada, T. Sata, "The Integration of CAD and CAM by Application of Artificial-Intelligence Techniques". Annals of the CIRP, Vol. 31/1/82.

4.   Descotte, Y., and Latombe, J.C., "GARI: A Problem Solver that Plans How to Machine Mechanical Parts," Proceedings 8th IJCAI, Vancouver, Canada, 1981.

5.   Gini, Giuseppina, Maria Gini, Rosamaria Morpungo, "A Knowledge-Based Consultation System for Automatic Maintenance and Repair." Proceedings 5th IFIP/IFAC Conference on Programming Research and Operations Logistics in Advanced Manufacturing Technology, Leningrad, USSR, 1982.

6.   Zdeblick, William J., "Computer-Aided Process and Operation Planning", Realizing Your Factory of the Future - Today, Cincinnati, Ohio: Metcut Research Associates, 1983, pp. 5-1 to 5-19.

7. Ackenhausen, A. F. and Harvey, S.M., _Machining Briefs_, Machinability Data Center, Cincinnati, OH, 1984, No. 2.

8. Preiss, K., E. Kaplansky, "Solving CAD/CAM Problems by Heuristic Progamming". _Computers in Mechanical Engineering_, September, 1983.

9. _Machining Data Handbook_, 3rd edition, Cincinnati, OH: Metcut Research Associates, Inc., 1980.

10. Balakrishnan, P., M. F. DeVries, "A Review of Computerized Machinability Data Base Systems." Tenth North American Manufacturing Research Conference Proceedings, Ontario, Canada: Society of Manufacturing Engineers, 1982, pp. 348-356.

11. Fisher, Edward L., Shimon, Y. Nof, Andrew B. Whinston, "Data Retrieval for Expert Systems," Computer Integrated Design, Manufacturing and Automation Center, Productivity Reports No. 5, 1984. West Lafayette, Indiana. Potter Engineering Center, Purdue University.

12. Adlard, Edward J., "The Use of Flexible Group Technology Codes in Process Planning", Proceedings 19th Annual Meeting and Technical Conference, Numerical Control Society, Dearborn, Michigan, 1982.

13. Zdeblick, W. J., and DeVor, R. E., "A Comprehensive Machining Cost Model and Optimization Technique," Annals of the CIRP, Vol. 30/1/1981, pp. 405-408.

14. Zdeblick, W. J., "An Adaptive Planning Methodology for Machining Operations," Proceedings SME International Tool & Manufacturing Engineering Conference, Philadelphia, Pennsylvania, 1982.

15. Zdeblick, W. J., L. J. Hawkins, Jeff R. Lindberg, "Machinability Data Base for End Mill Application." International Tool & Manufacturing Engineering Conference, Detroit, Michigan: Society of Manufacturing Engineers, 1981.

16. Lindberg, Jeff R., Sadayuki Nakamura, William J. Zdeblick, "An Integrated Machining Data Selection System". Tenth North American Manufacturing Research Conference Proceedings, Ontario, Canada: Society of Manufacturing Engineers, 1982, pp. 357-363.

17. Hayes-Roth, Frederick, Donald A. Waterman, Douglas B. Lenat (eds.), _Building Expert Systems_, Addision Wesley, Reading, Massachusetts, 1983.

18. Nau, Dana S., "Expert Computer Systems and Their Applicability to Automated Manufacturing," U.S. Department of Commerce, National Bureau of Standards, NBSlR 81-2466, Washington, D. C., 1982.

Automated Part Programming for CNC Milling by Artificial Intelligence Techniques

K. Preiss, Ben Gurion University of the Negev, Beer Sheva, Israel
E. Kaplansky, Man-Machine Cooperation Ltd., Beer Sheva, Israel

Abstract

This paper discusses research which has led to a working program based on artificial intelligence techniques for automatically writing a part program for milling. The paper discusses the approach used and gives details of the implementation. The input to the program is the graphic representation of the part (a drawing), and user-defined items such as tool details, material type, and so on. The program has an initial state, the shape of the raw material, and a goal state, the shape of the part. The program solves the problem of achieving the goal state from the initial state by using machining moves, and hence writes the part program. The current implementation produces a part program for a 2 1/2-dimensional part on a 3-axis CNC milling machine.

Keywords: Artificial Intelligence, CAD, CAM, CNC, Milling, Machining.

Research has been carried out over the past decade to try and discover how the human cognitive process works. This research has included the efforts of many people covering a spectrum of disciplines from psychology to logic and computers. Flowing from that work has come some understanding of how the thinking process occurs in the brain, and those principles are being applied to various problems in different fields. This paper is devoted to the application of artificial intelligence (AI) tech-niques to the problem of automatically writing a part program for milling. The paper will discuss not only the principles, but also their implementation into a working program. Implementation details are no less important than the theory.

The aim of a program such as this presently seems pretentious, in that we wish to replace human decision making by a program. That program will have to solve not only the usual and simple cases of machining, but also all kinds of special cases which today are taken care of by the human part pro-grammer. For instance, many part programming systems cannot manage to deal with edges which meet tangentially, and the human user can use one of several "fixes" to overcome difficulties caused by such a case. Or, if faced with an unusually shaped pocket, a user will usually divide the pocket into several pockets which he can manage. To deal with such cases, the computer will have to decide how to split the pocket, and split in the optimal manner.

The program will have to have encoded instruc-tions which represent knowledge about milling, and with all the cases that can occur. The procedures of the program (discussed later) will have to take account of tangential meetings of edges, short edges, concave pockets, islands in pockets, regions too small for the cutter, and many other cases that can occur. Also, if a case occurs that the knowledge in the program cannot manage, the program should not fail, but should send a detailed error message and should gracefully and systematically stop.

Because the knowledge of machining has been incorporated into the program, the program must have two essential human engineering characteristics. Firstly, it must not make decisions for a human, but must present its output as a recommendation to the user, which the user may accept or change. If this property were missing from a system, it would not be used. It may be that at some future time, a person responsible for metal cutting would trust a part program written by a computer, but for the forseeable future, that will not be the case. It is essential that the person in charge know what the computer is doing, and edit or accept the program.

Secondly, the system must be capable of operation by an operator at a low technical level. Because the system will be able to write a part program, the expert part programmer who knows both machining and programming should not have to spend his valuable time operating a laborious system. The system should be operated by a low technical level operator who receives a copy of a drawing and various data from the expert, inputs the information, generates the part program, and displays it on a simulation for the expert to edit and accept. If an artificial intelligence program were not organized in such a way, it is unlikely that it would have any impact on the machine shop. Because of the importance of these human engineering factors, this paper will discuss both the theory of the method and the implementation.

It should be emphasized that the problem dealt with here is different from the problem of process control usually dealt with in the literature. There are papers dealing with the problem of process planning—which machines to use to produce a part, what technological parameters to use—and so on. For the problem addressed here, the programmer writes the part program in the context of a technology defined by a user. The current implementation does not do any heuristic search or artificial intelligence processing of the technological parameters. The rules of the technology in the program are the handbook formulae, and the user specifies which machine and which tools and fixtures to use.

Structure of an AI Program

In the usual kind of computer program, the user decides how to solve a problem and the program performs the computational work according to the encoded instructions. In the new and growing class of programs structured according to the principles of artificial intelligence, the program decides how to solve a problem, and then shows the solution. An artificial intelligence program can be written in any computer language, although LISP and PROLOG are often used. The fact that an artificial intelligence program is a series of definite instructions like any other program, should not obscure the fact that the structure of an AI program is different from a deterministic program. The concept behind such programs is that of a state of the data and of operators which convert one state to another in order to reach a goal state. A state, in this case, is the data defining the state of the material and the machine tool at any general instant during the machining process. An operator is a machining move which changes one state of the data to another.

In order to solve a problem, one needs a definition of an initial state and of a goal state. One also needs a sufficient collection of operators to be able to convert the initial state through many intermediate states to the goal state. Also, and very importantly, one needs a logical structure which will decide in which order to try the operators in order to solve the problem in a reasonable computation time. At any instant, the order of trying the operators is itself dependant on the then current state of the data. As a result, an important characteristic of a search program which can find a solution is the ability to backtrack. The program may apply operators in such a way that a current state for which there is no route to the goal is reached, and then a new route is chosen. In general, the aim of the control structure is to solve the problem with as little backtracking as possible. Furthermore, it is common to define an evaluation function, which represents a weighted "cost" of the solution path, and then to search for the least cost solution.

Programs with such a structure have been written for many fields of activity such as medical diagnosis, geological prospecting, designing computer configurations, interpreting pictures, interpreting natural language, writing part programs, planning production, and much else. These programs have lately been the focus of growing interest and review.[1-4]

Problem Solving

Consider a program which plays chess. In such a program the current state is the state of the chess board, and the goal state is the elimination of the opposing king. The operators are the moves of chess. One can ask the question "Can I win this game?", and can then let the operators work until they come up with a yes or no answer. Then one can ask for a proof of the answer and the computer will display the moves which lead to the goal state. Note that in order that the solution be generally true, the proof is done by assuming the negation of the statement to be proved, and then seeing if that negated statement is a logical contradiction. If it is, then the statement is true and the ordered application of the operators is a proof of the statement.

It is sometimes erroneously thought that the computer solves such a problem by evaluating all the implications of all the moves possible from a given state. For the game of chess, it has been calculated that such an evaluation would require, on the average, a century of computing and is hence impractical. The problem is solved by various techniques, especially the use of knowledge or heuristic information to reduce searching in unproductive directions and to concentrate on more fruitful directions.

In the problem mentioned above, each state of the chess board considered during the calculation is represented by a string of characters. The initial state and the final state are also represented by a string of characters, therefore, the problem can be considered as one of mathematical grammar where one wants to prove that the final string is achievable from the initial string by applying the rules of the grammar.

There are various approaches to a theorem proving problem. An elegant method is that of resolution,[5] another method is by the use of condition-action rules, often called production rules. The basis of the resolution method is the fact that the following logical statements are equivalent:

IF (A is true) THEN (B is true) is equivalent to
.NOT.(A is true).OR.(B is true)

and

(.NOT.(A is true).OR.(B is true)).AND.((A is true).OR.(C is true)) can be replaced by ((B is true).OR.(C is true))

The IF A THEN B pairs are converted to the .NOT.A.OR.B form, and the many clauses are then fused or resolved, one at a time, until it is found that the negated statement is both true and false, which is a contradiction, so the statement is true. The computational problem here is to rapidly decide at each point in the problem which are the two clauses to resolve so as to obtain the solution.

The work presented in this paper was not done by the resolution method, but by directly using the condition-action rules. The knowledge about milling is embedded in those rules. To solve a problem, one applies the rules until the goal state is achieved. One cannot simply try each rule in a fixed order at each state, because that process would take too long. The order of trying the rules has to be dynamically controlled according to the state of the data, and also, rules are grouped together in a hierarchic structure. This enables one to disregard whole groups of rules which are not applicable at a given state without even evaluating the preconditions of those rules.

Automatic Part Programming

Before discussing the current problem, let us discuss automatic part programming methods for 2-D problems. Flame cutting and simple turning are examples of 2-D problems; for flame cutting the problem is in the X-Y plane, for turning it is in the R-X plane, R being the radius of the material and X the axis of the lathe. In each of these problems, a single strategy of machining will solve all problems (see *Figure 1*). Common strategies can be described as follows:

For flame cutting:
> Repeat
>> Go to a profile curve, which is usually a closed loop of edges.
>> Repeat
>>> Cut along an edge.
>> Until the loop of edges closes.
> Until all the loops are done.

For simple turning without overhang:
> Set the tool at the clearance height in direction $+R$.
> Repeat
>> Move to the start value of X.
>> Repeat
>>> Move the tool along the X direction until it is at a region to be removed.

Move the $-R$ direction until a specified depth of cut is achieved.
Move in the $+X$ and R directions, cutting material.
Until the tool gets to the end of the material in the X direction.
Until the part is cut.

For 3-axis milling, many systems use the following strategy:

Set the cutter at a corner of the raw material.
Repeat
    Raise the cutter to the clearance plane, clear of the material.
    Move the cutter a small distance $+Y$.
    Repeat
        Raise the cutter to the clearance plane.
        Move in the X direction until at the beginning of a region needing removal.
        Move down in the Z direction until a specified depth of cut is reached.
        Cut in the X direction with that depth of cut, up to the part boundary.
    Until the part has been traversed along the X direction.
Until the part has been covered in the Y direction.

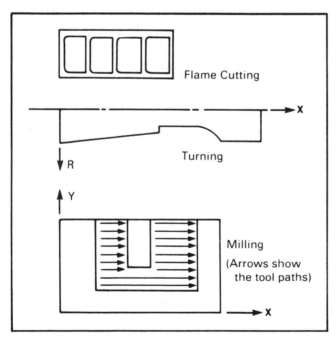

Figure 1
Geometries for Common
2-D Automatic Machining Methods

In this solution, the 3-D milling problem is decomposed into many 2-D problems in series with loss of optimality in the solution. This strategy can indeed manage pocketing, but with much machining time and sometimes with moves which are not

considered good technique. The solutions mentioned above are each with a permanent strategy, which is sufficient for a 2-dimensional problem.

All three of the above algorithms can be written as follows:

Set the tool at a start location, and a safe height.
Repeat
    Move the tool a small distance in the secondary direction.
    Repeat
        Move to an entity (such as a loop of edges or a pocket) which needs cutting.
        Repeat
            Cut in the major direction.
        Until the entity (such as the loop of edges or the section of the pocket) is traversed.
    Until there are no more entities to do in the current major direction.
Until the material is all covered and done.

After these introductory remarks, we can describe the automatic CNC milling part programming problem.

Given:

   a. An initial state (the raw material).
   b. A goal state (the part).
   c. A set of operators (milling machine operations).
   d. A control structure which decides which operation to do next.

Required:

A proof that the operators can construct the part from the raw material. We wish the proof to be with the least cost, that is, with the least value of the evaluation function.

Each move of the proof will be a machining operation, and the requirement for a minimum cost proof will lead to an optimum part program. It may be noted that a proof can be derived by operating from the initial state to the goal state (forward chaining) or by starting with the goal state, using reversed operators, and seeing if the initial state is reachable (backward chaining).

We may note that for a 2-D problem such as flame cutting, there is a single proof, always true and optimal. It is a tautology, or a universally true statement, as follows.

    Assertion: A part is defined by a path of non-intersecting edges, with defined start point and end point.
    Operator: Cut along an edge.

Control Start at the start point and cut
Structure: along edges until the end point is
reached, or until it cannot be reached.

The only case for which the end point will not be reached is when there is a fault in the definition of the part, and so the algorithm must terminate. The other 2-D problems are solved by a similar strategy. It is a trivial exercise to take a situation defined to be true, then to ask for a proof of that assertion using a permanent series of operators. Hence, 2-D problems are deterministic and do not need theorem proving methods.

For milling a part with an arbitrary perimeter and many pockets and drill holes using many different possible tool paths and machining operations, there are many possible series of operators which can produce the part, and hence, many proofs of the theorem. Therefore, for this problem one requires a program which will find the optimal proof.

Overview of the Solution

When having to write a stream of instructions for machining a part, the initial state is the data defining the shape of the raw material, the tool type and dimensions, and the tool location. The final state is the shape of the part with its tolerances and finishes. At any intermediate stage, the state is defined by the then current shape of the part, the tool type and size at that stage, the tool location, the current tool speed, and the then current status of other machine variables.

Definition of the intermediate state includes the state of the computation at that state. The computation may be driven through many states which are not on the path to the goal, but which are evaluated and later rejected. The aim of a well written program is of course, to evaluate the path to the goal state with as little as possible computation of states, which later are found not to be on the path to the goal.

The operators which will change one state to another are machining actions. *Figure 2* shows suitable operators for 3-axis milling; a brief explanation is given in *Table 1*. Note that the operators are arranged in hierarchy. Each operator is called by a higher level operator and, except for the lowest

level, calls on lower level operators. The lowest level operator, when called, generates an appropriate CNC instruction and simultaneously changes the state of the data in accordance with the action implied by the operator. Each operator has attached to it a set of preconditions, as a single logical function, which is evaluated before the operator is invoked.

The structure of the operator is:

if <CONDITION> then <ACTION>

This structure is known in the literature as a production rule, or as a condition-action pair. If a rule cannot be applied because its condition is false, then the next applicable rule is tried. If all the rules at a level cannot be tried, the program backs up to the higher level, and so on, recursively. At each stage of the computation, the program tries to apply the operators in an order selected by the control mechanism. The precondition of each operator is evaluated and the first operator which can be applied, is applied. If, at any level, no operator is applicable, the program reverts to the next operator at the level above. If an operator can be applied, the operators called by it, at the next level, are called. If the control backs up to the root node, the problem is unsolvable. A number of flags, or logical variables, keep track of the states of the computation and of the reasons that a condition of a rule is found to be false. Hence, as control backs up the tree of operators, information about the state of the data space flows up the tree. This information can be used to reorder the use of operators by deciding, according to the state of the computation, what to do next.

Consider, for instance, the case of a cutter with a diameter too big for a pocket. If it is so big that not even one cut can be made, control goes to the change tool operator. If the tool is not changed, control is passed up to the start node and the program will stop with a message that the part cannot be made. If, on the other hand, the diameter is such that some pocketing can be done, but there is a narrow region through which the tool cannot move, then the program provides the user with the option of changing the tool, in which case control passes to the change tool operator and then continues as usual, or else (if he does not wish to change the tool), the control will not move away from the take layer operator, and the program will cut what it can.

Note that in this case, the program will search

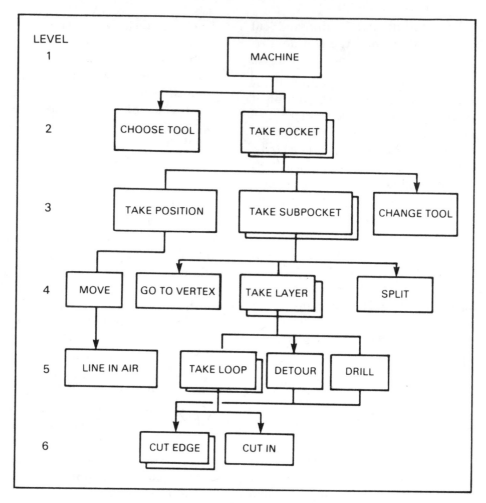

Figure 2
The Operators
(Operators at any level are tried from left to right. A double box indicates
that an operator is repeatedly tried while its preconditions hold.)

for the optimal proof that the part will be cut. The program will notify the user that the theorem cannot be proved, and hence the part not cut, but will present the user with the steps of the proof as it tried to do it. This list of steps is the part program. Because the preconditions to the operators include checks for collisions with the part and with clamps, the part program will cut whatever can be done, but without damaging the part or clamps.

The program is made more efficient by conducting the search from the final state, that is from the part, to the initial state, that is, the raw material. If one would want to choose the operators by considering how to get from the raw material to the part, then one would have to perform a 3-dimensional pattern recognition activity so as to evaluate the order of applying operators and the precondition for each operator. That pattern recognition activity would be similar to the decisions of allocating group classifications in group technology.

Having adopted the approach explained here, several other basic questions fall in place. The cutting tool will always run along an existing edge beside a face. Hence, the boundary definition method of representing the part[6] is suitable. It also becomes unnecessary to identify pockets before invoking the operators. As soon as the topology of the current part state changes, then a pocket has been completed.

Among the operators provided, there is one which changes the state space but is not a machining operator, and that is the split action. That changes the data structure by splitting a pocket into two so

The operators marked * create CNC instructions.

MACHINE	The Main Procedure.
CHOOSE TOOL*	Choose a tool.
TAKE POCKET	Deal with a pocket.
TAKE POSITION	Go the the position from which the pocket will be solved.
MOVE	Go from point to point not cutting.
LINE IN AIR*	Make a straight line move.
TAKE SUBPOCKET	Deal with a subpocket.
GO TO VERTEX*	Go to a vertex from which to work.
SPLIT	Change the data structure from one pocket to two.
CHANGE TOOL*	Change the tool.
TAKE LAYER	Deal with a layer of a pocket.
TAKE LOOP	Solve the problem for a loop of edges.
DRILL*	Move into a layer to be cut.
DETOUR	Move around an obstacle.
CUT EDGE*	Cut along an edge.
CUTIN*	The first move into a loop.

that the program can continue to work to solve the problem, but it does not lead directly to any machine action.

The operators chosen by heuristic search are usually chosen by making the best choice according to some evaluation function.[5] In the program written here, an evaluation function was not explicitly computed at each step. However, the control structure and preconditions attached to each operator were written so as to provide an action which would give an optimum value of the cost of the solution. It is known[5] that if the predicted cost of the solution is below a calculable lower bound for the problem, then the solution will be optimal.

For machining, the time required to finish a part must be more than the time which would be required if the tool only cut metal, and did not move in air. Hence, for given feed and speed, the function found by taking the volume of the remaining metal to be removed will be the basis for an evaluation function which assures optimality. If, at any decision point, the operator is chosen which will remove the most metal commensurate with the technological data, then that choice is implicitly derived from an evaluation function less that the lower practical cost and must give an optimal solution.

For 3-D milling as a series of 2-D problems mentioned in the previous section, the direction of cut is not chosen by the control structure and the operators, but is defined by the strategy of the operations without reference to the current state of the data. The 2-D solution is hence not optimal. Only in coincidental particular cases will the 2-D strategy give the optimal cutting solution.

Implementation

The principles described above have been implemented in a program written in Fortran 77, for the 3-axis milling of 2 1/2-dimensional parts. The program writes a part program in EIA format (G codes). The part program gives the optimal tool paths to manufacture the part. If the problem is not solvable, for instance because the tool is too big for certain regions of the part, the program will identify such a situation and will not cut in a way which will ruin the part. The program manages situations which could lead to computational problems such as edges of length less than the tool radius, or two curved edges meeting contangentially. The part can consist of holes to drill, a profile, and pockets, and the program writes a series of instructions to convert the initial state (the raw material) to the final state (the part).

The program is embedded in the MINC (Machine Intelligence for Numerical Control) software system. The total system consists of 25,000 Fortran 77 statements in a number of files and modules. It is a portable package, having run on VAX 750, VAX 780, CDC Cyber NOS, Data General MV series and Cromemco System 1 computers, with Tektronix, Lexidata, Visual and Genisco graphic terminals.

The MINC system consists of a series of programs as shown in *Figure 3*:

1. DRAW is a menu-driven interactive graphics program used to input a drawing of the part. Input can also be achieved by reading a file defining a part from other drawing packages. The program converts the representation of the design drawing to a "clean" representation suitable for the part programming program. Multiply-defined lines near intersections, lines not part of a closed loop and other common occurrences in design drafting packages, are detected and taken care of automatically.

2. SOLID is a program which automatically

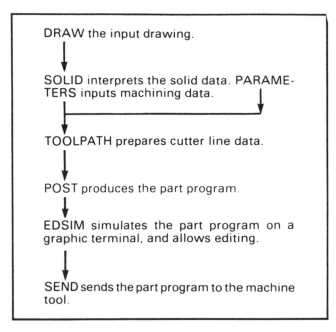

DRAW the input drawing.

SOLID interprets the solid data. PARAMETERS inputs machining data.

TOOLPATH prepares cutter line data.

POST produces the part program.

EDSIM simulates the part program on a graphic terminal, and allows editing.

SEND sends the part program to the machine tool.

Figure 3
The Programs of the System

interprets the solid part shape from the drawing, using values of the Z dimension input by the user.

3. PARAMETERS is an interactive program to input various technical data including tool data such as diameter, corner radius, material, etc; tool motion, clockwise or counterclockwise; distance from the part contour to be left for finishing; home coordinates; clearance height; overlap, and so on.

   Speeds are chosen from tables encoded in the program. The user is shown the computed values and can override them, but if he chooses unreasonable values, they will not be accepted.

4. TOOLPATH prepares a cutter line part program. This automatically deals with pocket, profiles, and holes to be drilled. The contours can be any combination at all of straight lines and circles. Circles are true circles, and not straight line segments, and the program deals with the computational problems arising when many circles and straight lines meet tangentially. Any number of pockets can be dealt with, and they may be entirely within the material, or may cut into a side or corner. The program deals with computational problems caused by edges of length less than the radius of

the cutter, and will machine exactly to the corner of the short edge. Regions which may be left at sharp corners, due to nonoverlapping paths, are automatically cleaned out. The tool path will not cut into narrow regions or small bays for which the tool is too large.

5. POST produces the part program, which is currently in EIA (G code) format for several CNC machines which use that language. The part program is liberally provided with comments inserted by the system. As a result, the part program is comprehensible not only to experts in the part programming language, but also to lesser experts. *Figure 4* gives an example of a portion of a part program.

6. SIMULATE is a program which accepts the part program and drawing of the parts as input and shows the machining moves as a simulation on a graphic screen. As the simulation progresses, the part program scrolls along on the screen so the user can see which line of part program produced which cutting move, making it easier for him to edit the part program if he wishes. The part program can be edited and checked on the simulation before being approved and released.

7. SEND sends the part program to the appropriate device, which may be paper tape, a controller, or whatever. The user specifies if he wants it dispatched with or without comments.

Using the MINC system, a part program can be totally prepared in less than half an hour for a simple part. After the input is defined, a few minutes of computer time is needed to produce the part program. Furthermore, the work of inputting the drawing and data, producing the part program and bringing it to the simulator, can be done by an operator level technician. The expensive experienced part programmer is saved the time of operating the computer system and can devote his attention to the simulation and approval.

Initial experience using the program in an industry has shown that users are happy to be relieved of the intellectual burden of figuring out where the machine should cut, but wish to retain control of the strategy, the order of machining, and the like. Development effort has therefore been devoted to making the system less automatic than it could be.

```
N100 (*   PART PROGRAM FOR FANUC MACHINE NO. 2. *)
N101 G40 G80 G90 ;
N102 G49 G58 M23 ;
N103 T22  ;
N104 (*   LOAD T22, SET T09 *) ;
N105 T09 ;
N120 (*   BEGIN job with  1 PROFILES and  1 POCKETS. *)
N130 (*   START working with TOOL no.: 22, tool diameter = 10.00 mm. *)
N140 (*   START moving to the first POCKET. *)
N160 G1 F2500 X76.937 Y62.5;
N161 G43 H22 Z10.;
N162 M03 S2000;
N163 M08;
N170 Z1.;
N180 (*   COMPLETED moving to the first POCKET. *)
N190 (*   Start cutting SET-UP no.: 1. *)
N200 (*   START taking POCKET no.: 1            , in SET-UP no.: 1. *)
N210 (*   START taking SUB_POCKET no.: 1, in POCKET no.: 1. *)
N220 (*   START cutting down into LAYER no.: 1, in SUB_POCKET no.: 1. *)
N230 F30 Z-6.5;
N240 (*   COMPLETED cutting down into LAYER no.: 1, in SUB_POCKET no.: 1. *)
N250 (*   START taking LAYER no.: 1 in SUB_POCKET no.: 1. *)
N260 (*   START taking inner rough cuts in SUB-LAYER no.: 1, in LAYER no.: 1. *)
N270 (*   START taking LOOP no.: 1, in LAYER no.: 1. *)
N280 F40 X86.682 Y67.338;
N290 G2 X86.682 Y57.663 I-36.682 J-4.838;
N300 G1 X76.937 Y62.5;
N310 (*   COMPLETED taking LOOP no.: 1 in LAYER no.: 1. *)
N320 (*   START cut_in into LOOP NO.: 2. *)
N330 F30 X69.797 Y60.462;
N340 (*   COMPLETED cutting-in into LOOP no.: 2. *)
N350 (*   START taking LOOP no.: 2, in LAYER no.: 1. *)
N360 F40 Y64.538;
N370 X90.199 Y74.666;
N380 G2 X90.2 Y50.334 I-40.199 J-12.166;
N390 G1 X69.797 Y60.462;
N400 (*   COMPLETED taking LOOP no.: 2 in LAYER no.: 1. *)
N410 (*   START cut_in into LOOP NO.: 3. *)
N420 F30 X64.797 Y57.362;
N430 (*   COMPLETED cutting-in into LOOP no.: 3. *)
N440 (*   START taking LOOP no.: 3, in LAYER no.: 1. *)
N450 (*   START taking the last LOOP in LAYER no.: 1. *)
N460 F40 Y67.638;
N470 X92.94 Y81.608;
N480 G2 X92.94 Y43.392 I-42.94 J-19.108;
N490 G1 X64.797 Y57.362;
N500 (*   COMPLETED taking the last LOOP in LAYER no.: 1. *)
N510 (*   START moving to PROFILE no.: 1. *)
N520 F2500 Z10.;
N530 X0. Y44.;
N540 Z1.;
N550 (*   COMPLETED moving to PROFILE no.: 1. *)
N560 (*   COMPLETED taking LAYER no.: 1 in SUB_POCKET no.: 1. *)
N570 (*   COMPLETED SUB_POCKET no.: 1, in POCKET no.: 1. *)
N580 (*   COMPLETED POCKET no.: 1, in SET_UP no.: 1. *)
N590 (*   START taking PROFILE no.: 1, in SET-UP no.: 1. *)
N600 (*   START cutting down into LAYER no.: 1, in PROFILE no.: 1. *)
N610 F30 Z-6.5;
N620 (*   COMPLETED cutting down into LAYER NO.: 1, in PROFILE no.: 1. *)
N630 (*   START taking LAYER no.: 1 in PROFILE no. 1. *)
N640 (*   START taking the last LOOP in LAYER no.: 1. *)
N650 G2 F40 X-6. Y50. I0. J6.;
N660 G1 X-6. Y75.;
```

Figure 4

A Portion of a Part Program Automatically Written by the System (Note the comments which enable the user to understand the program.)

Input and Output Interfaces

This collection of programs has an input interface with the user and an output interface to a machine tool. The input interface has to be easy and convenient for the user, and this side of the package requires a detailed approach to the user interface as discussed in the literature for any interactive package. The output interface has to produce part programs comprehensible to the controller of the milling machine, and if even one character is wrong or misplaced, the milling machine may not function correctly.

It should be emphasized that for different controllers, even from the same manufacturer which nominally follow the same standard, there can be differences. In order to correctly control the system, and as much as possible, avoid error due to human interaction, the programs are included in a system shown in *Figure 3*, and the whole system is managed by a command program which presents the choices to the user in simple English. There are several principles adhered to throughout the system in all the programs:

1. When given the input, the program will output a part program. In principle, it is not desirable that the program make decisions for a human, so the output is presented as a suggestion which the human can change or accept. This leads to several aspects discussed below.

2. The user has no recourse to operating system commands. File management, configuration control, and keeping track of update times and currency of the drawings and part programs is handled by the command program, not by the user, so that the user is relieved of the annoying requirement that he must understand how to manage files and how to keep everything updated.

3. If a user inputs data with wrong syntax, or an impossible or out of range attribute, or asks to do a task known to be impossible, the system will print an explanatory error message and will not accept the unallowable data.

4. Whenever possible, the computer will suggest a default answer for input. By pushing <return> on the keyboard, that default choice will be made. The user can then rely on the knowledge encoded in the computer, or can override it. For instance, when specifying machining data, the user can ask for the values from a permanent file, or the values from a previous run, together with formulae encoded in the computer. He then runs through the questions and for each question can use the data suggested by the computer, or can input his own desired data.

5. The part program includes, in addition to the lines of CNC code, comments in English which explain what is being done in the part program. A user not specially expert in part programming can then read and understand the documented part program (see *Figure 4*).

6. The part program is presented on a graphic simulator for editing and approval. The part program as suggested by the system, is a program which has been written and documented by the computer program. It will therefore not have any careless errors (if there were errors they would be due to incomplete rules and conditions in the program that wrote it). If anyone changes that part program, he must bear the responsibility for any errors he may introduce.

7. The above points refer to the interface to the human user. There is a second interface from the system to the CNC milling machine controller. This interface must of course produce the correct desired actions at that machine. That is indeed the case, as is proved by the success achieved in cutting real parts.

Examples

The MINC system permits the user to carry out the whole process of part programming in a single system, while relieving the user of the task of computing the tool paths, or of figuring out where the mill should move to cut a part. This dynamic process is difficult to illustrate in figures, just as it is difficult to give the impression of a movie from some still pictures.

Figures 5, 6, 7, 8, and *9* illustrate simulations of part programs automatically created by the system. These examples show the following characteristics:

1. The whole part program is automatically written in a part program which includes explanatory comments.

2. Edges which meet tangentially at 180° are taken care of.

3. Edges of length less than the tool radius cause no problem.

4. The tool will cut only in those regions where the part is not damaged, and then will cut exactly as much as can be cut according to the geometry of the problem.

5. The tool cuts include short forays into regions of narrow angle, which if they were left, would leave an uncut patch in between the paths.

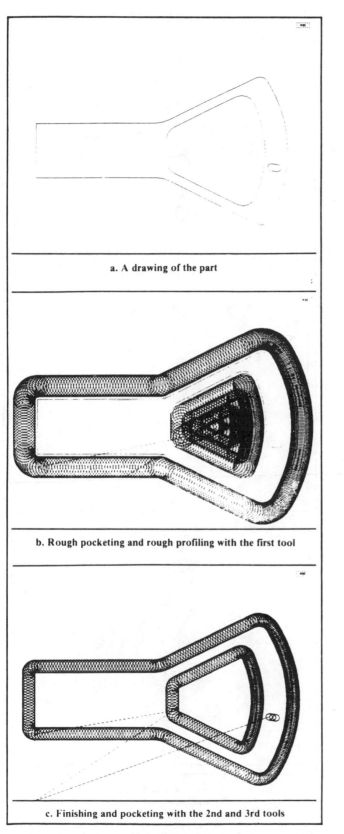

a. A drawing of the part

b. Rough pocketing and rough profiling with the first tool

c. Finishing and pocketing with the 2nd and 3rd tools

Figure 5
Stages in Milling an Example Part

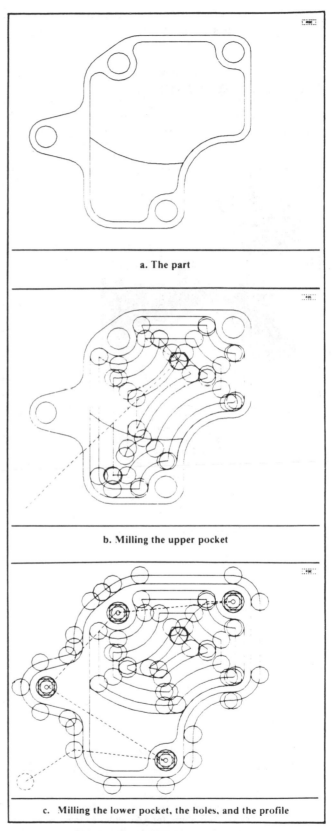

a. The part

b. Milling the upper pocket

c. Milling the lower pocket, the holes, and the profile

Figure 6
Stages in Milling an Example Part

6. The tool path is such that the cutting procedure spends the minimum time in air, and hence is optimal.

Concluding Remarks

Knowledge of milling has been successfully encoded into a computer program using principles

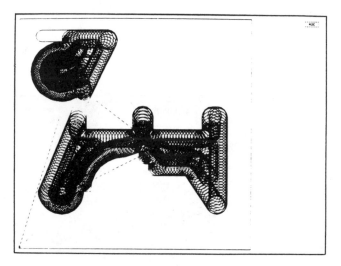

Figure 7
**Example of an Automatically
Chosen Tool Paths for a Complex Pocket**
(Moves in air are shown as dashed lines. The full tool is shown
on the simulation for clarity.)

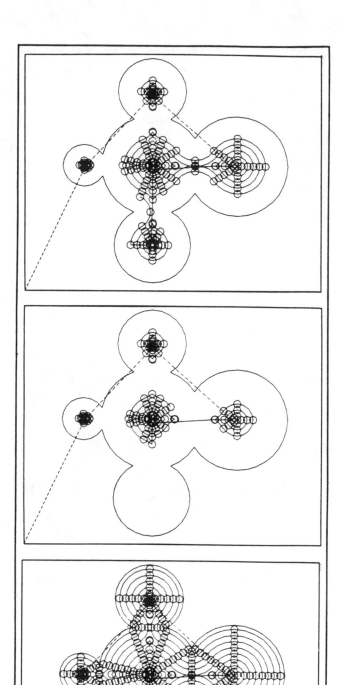

Figure 8
**Example of an Automatically
Chosen Tool Paths for a Complex Pocket**
(A complex pocket with a tool small enough to cut the entire pocket.
Moves in air are shown as dashed lines. The tool paths are shown with
the tool shown in full at the ends of each move.)

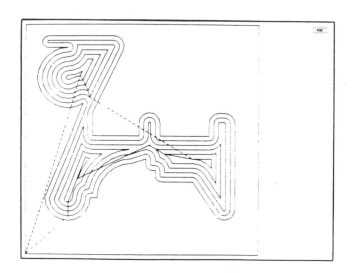

Figure 9
**An Example of an Automatically
Chosen Tool Paths for a Complex Pocket**
(A complex pocket with a tool small enough to cut the entire pocket.
Moves in air are shown as dashed lines. The tool paths are shown.)

of artificial intelligence, and a system has been produced which automatically writes part programs.

The system written has been used to successfully mill 2 1/2-dimensional parts on 3-axis milling machines. The principles, as described here, can be exerted to 3 and more axis milling. The package will also deal, in principle, with 2-dimensional machining, such as flame cutting or turning. As previously explained, the principles encoded into the program will solve such 2-dimensional problems as trivial cases of the 3-dimensional milling.

References

1. J.R. Dixon, M.K. Simmons. "Computers That Design: Expert Systems for Mechanical Engineers", *Computers in Mechanical Engineering*, ASME, Volume 2, No. 3, November 1983, pp. 10-18.
2. W.B. Gevarter. "The Languages and Computers of Artificial Intelligence", *Computers in Mechanical Engineering*, ASME, Volume 2, No. 3, November 1983, pp. 33-38.
3. P. Kinnucan. "Computers That Think Like Experts", *High Technology*, Volume 4, No. 1, January 1984, pp. 30-42.
4. K. Preiss, E. Kaplansky. "Solving CAD-CAM Problems by Heuristic Methods", *Computers in Mechanical Engineering*, ASME, Volume 2, No. 2, November 1983, pp. 56-60.
5. N.J. Nillson. *Problem-Solving Methods in Artificial Intelligence*, McGraw Hill, 1971.
6. C.M. Eastman, K. Preiss. "A Review of Solid Shape Modelling Based on Integrity Verification", *Computer Aided Design*, Volume 16, No. 2, March 1984, pp. 66-80.

Author(s) Biography

Kenneth Preiss is a Sir Leon Bagrit Professor of CAD in the Department of Mechanical Engineering at the Ben-Gurion University of the Negev in Beer-Sheva, Israel. He obtained his Ph.D. in Nuclear Engineering at the Imperial College of Science and Technology in London in 1964. Professor Preiss has taught at the University of Illinois, Urbana, and since 1966 has been at Ben-Gurion University.

Professor Preiss' rescarch interests are in artificial intelligence, solid modelling, the CAD/CAM process, and related fields. He is a member of several international organizations, including SME, and is an honorary member of ASME.

Eliezer Kaplansky earned a B.S. degree (1980) in Mechanical Engineering and a M.S. degree (1982) from the Ben-Gurion University. The subject of his thesis was "Automatic Mill Routing from Solid Geometry Information". Mr. Kaplansky subsequently founded the Man-Machine Cooperation Ltd. with Professor Preiss where he has since been working.

Presented at the CASA/SME AUTOFACT 5 Conference, November 1985

Intelligent Manufacturing Planning Systems

by David Liu
Hughes Aircraft Company

Introduction

Manufacturing planning determines the overall sequence of operations, prescribes the step-by-step instructions of each operation, and schedules the operations. Many firms have developed computer-aided process planning (CAPP) systems to reduce the time required to sequence the operations. These systems can produce skeletal process plans. However, they still require a great deal of manual direction and input.

Today, some manufacturers are attempting to completely automate manufacturing planning. Using expert systems, an application of Artificial Intelligence, these firms are developing software that mimics the reasoning of manufacturing planners. Along with information from the engineering database, embodied facts, logic, and rules are applied to generate manufacturing plans automatically.

AUTOMATED PROCESS PLANNING SYSTEM

REQUIREMENTS

1. GENERATIVE

2. FLEXIBLE, PORTABLE

3. HUMANLIKE INTELLIGENCE

4. EASY TO USE

5. FREE PEOPLE FOR JOBS COMPUTER CANNOT HANDLE

6. INTEGRATE WITH REST OF MANUFACTURING SYSTEM

7. AVAILABLE EXPERTS

8. STABLE TASK DEFINITION

One of the first companies to implement such a system is Hughes Aircraft Company's Electro-Optical and Data Systems Group. The firm has developed the Hughes Integrated Classification Software System (HICLASS)^{TM} to integrate design and manufacturing. The objective is to electronically deliver computer generated planning throughout production. This enhances product quality, reduces support costs, increases flexibility and yields greater responsiveness.

To generate this on-line manufacturing planning, Hughes determined that the system must be capable of capturing manufacturing knowledge. The manufacturing knowledge is used to deduce required manufacturing operations from engineering databases. The design defines the problem, and available data includes graphical representation, engineering notes, design and manufacturing specifications, special manufacturing instructions, and inspection criteria.

HICLASS WELCOMES YOU TO THE REALM OF ARTIFICIAL INTELLIGENCE

Hughes made several attempts at developing such a system before settling on the current expert system. In 1982, the work began with a generative manufacturing planning system in which design features were matched to specific manufacturing operations. This system used so-called binary decision trees to produce process plans for printed circuit board (PCB) fabrication. Process networks and flows describing manufacturing operations were set up in this decision tree format. Although successful, the trees were too difficult to build. Moreover, process plannninging resembles an unstructured, 3D network, so mapping this network onto binary decision trees left out essential information and led to a loss of flexibility.

A second generation version was then built using multi-branch decision trees. Such trees were easier to build and modify. This system was benchmarked for three applications, including PCB fabrication, PCB assembly, and mechanical part fabrication. PCB applications proved that multi-branching worked well. Mechanical part applications, however, revealed the shortcomings of manually stepping through the decision trees. Planners would lose perspective of the problem after traversing a few levels into the decision trees.

HICLASS DEVELOPMENT HISTORY

1982 — PROTOTYPE PRINTED WIRING BOARD FABRICATION PLANNING SYSTEM USING DECISION TREES

1983 — PASCAL EXPERT SYSTEM PROGRAM WHICH CAPTURED LOGIC IN IF/THEN RULES

— DEMONSTRATION EXPERT SYSTEM WHICH CREATED PROCESS PLANS FOR 2 ASSEMBLIES

1984 — SYSTEM IN PRODUCTION FOR 2 ASSEMBLIES AT 5 WORK STATIONS

— BEGAN DEVELOPMENT OF "C" PROGRAM USING MORE COMPLEX IF/THEN/ELSE RULES AND ENCHANCED CONCEPTUAL STRUCTURES

This system, called the Hughes Integrated Classification Software System (HICLASS)^{TM}, used group technology software to classify parts and generate process plans. As an enhancement, software was included to infer part features based on part geometry. These inferred-part features guided the system in traversing decision trees and picking out essential manufacturing steps.

Inferring part features, however, proved to be a complex task. This early system required a significant amount of software and time to determine a route through decision trees. Such a structure was still not an optimal representation of manufacturing planning.

An entirely new system, also called HICLASS^{TM} Software System, taking the knowledge-based, expert systems approach was then developed. The system's capability to automatically extract part features from CAD frees planners from describing the part. This software deduces required manufacturing operations, instructions, and equipment programs from the engineering database through heuristics that represent manufacturing experience and intuition.

HICLASS CURRENT DEVELOPMENT

1985 — **TRANSFER ASSEMBLY PLANNING EXPERT SYSTEM TO NEW RULE FORMAT**

— **CREATE PLANNING FOR 32 ASSEMBLIES**

— **SYSTEM IN PRODUCTION AT 60 WORK STATIONS**

— **DEVELOP A GENERAL NATURAL LANGUAGE PARSER TO INTERPRET ENGINEERING DRAWING NOTES**

— **BEGIN DEVELOPMENT OF PROTOTYPE PRODUCIBILITY ANALYSIS SYSTEM FOR PRISMATIC PARTS**

The software operates in three phases to accomplish automated manufacturing planning. First, an interpretation phase defines the problem by translating engineering databases into "tokens" that can be processed as symbols rather than numbers. Next, a reasoning phase solves the planning problem by evaluating constraints and goals. Here, the symbolic processing continues until a conclusion or conclusions can be drawn. Finally, a presentation phase translates computer graphics which are sent to terminals on the shop floor. In the future, the presentation phase will also translate results into equipment programs that control the machinery.

HICLASS SYSTEM FUNCTIONAL DESCRIPTION

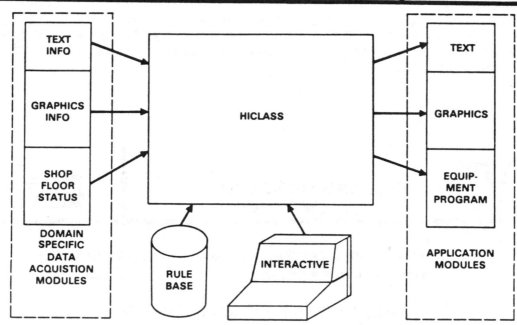

CURRENT CONFIGURATION

COMPUTER SYSTEM	APOLLO ENGINEERING WORK STATIONS
TERMINALS (5)	TEKTRONIX, COLOR
OPERATING SYSTEM	UNIX
LANGUAGE	PASCAL, C
NETWORK	APOLLO DOMAIN
INTERFACE	HP3000, CV/CALMA, IBM

Design engineers use interactive graphics to capture product descriptions into the engineering database, including geometric models, hidden-part attributes, and bills-of material. Also included are engineering notes, which contain part specifications, special instructions, and general design comments.

TRANSFER OF CAD DATA TO MANUFACTURING

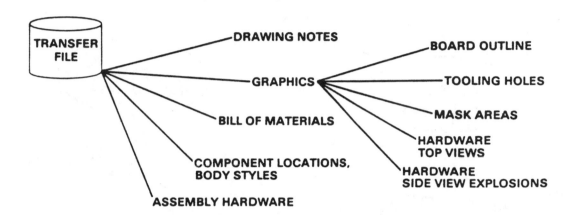

A complete product description is essential for correct manufacturing planning. If, for example, a "heat treat" note is missing, the resulting process plan may be incomplete.

Another facet of Artificial Intelligence known as natural language processing helps interpret engineering notes for use in manufacturing planning. The system uses a mixture of syntax parsing and semantics interpretation methods to understand these notes. Syntax provides grammar and semantics provides meaning. Both are represented by networks. Nodes in these networks represent word categories such as verb or noun, while arcs denote relationships between node (word or phrase) classes. Semantic interpretation provides meaning to words and phrases extracted from the engineering notes.

NATURAL LANGUAGE PROCESSING

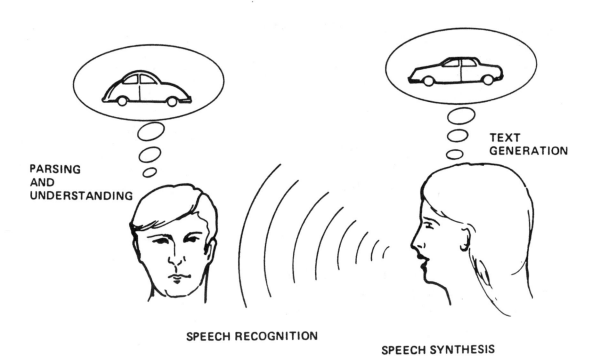

PARSING
AND
UNDERSTANDING

TEXT
GENERATION

SPEECH RECOGNITION

SPEECH SYNTHESIS

PHRASE STRUCTURE GRAMMER

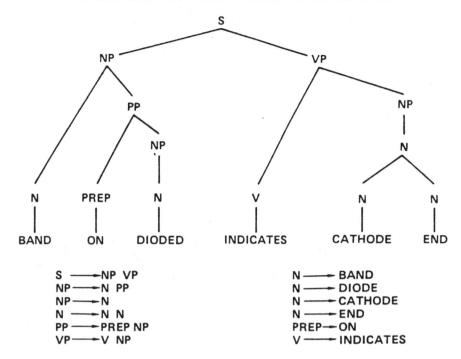

```
S  ———► NP VP          N ———► BAND
NP ———► N PP           N ———► DIODE
NP ———► N              N ———► CATHODE
N  ———► N N            N ———► END
PP ———► PREP NP        PREP ► ON
VP ———► V NP           V  ———► INDICATES
```

BOTTOM-UP PARSE STRATEGY

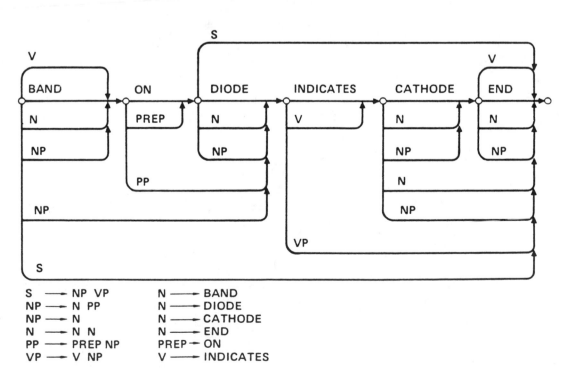

```
S  ———► NP VP          N ———► BAND
NP ———► N PP           N ———► DIODE
NP ———► N              N ———► CATHODE
N  ———► N N            N ———► END
PP ———► PREP NP        PREP ► ON
VP ———► V NP           V  ———► INDICATES
```

HICLASS NATURAL LANGUAGE
SUBSYSTEM

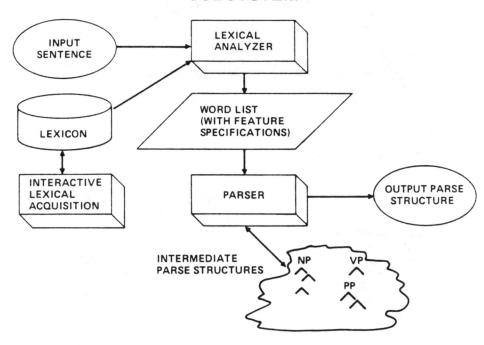

AUGMENTED TRANSITION
NETWORKS

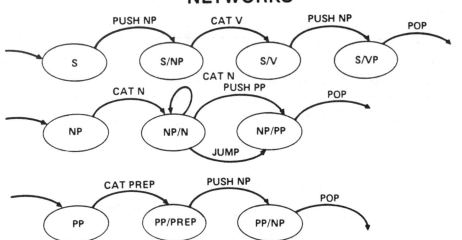

INPUT SENTENCE: BAND ON DIODE INDICATES CATHODE END

	STATE	ARC	CURRENT WORD		STATE	ARC	CURRENT WORD		STATE	ARC	CURRENT WORD
								15.	NP	CAT N	CATHODE
1.	S	PUSH NP	BAND	7.	NP	CAT N	DIODE	16.	NP/N	CAT N	END
2.	NP	CAT N	BAND	8.	NP/N	CAT N	INDICATES	17.	NP/N	CAT N	
3.	NP/N	CAT N	ON	9., 10., 11.				18., 19., 20.			
4.	NP/N	PUSH PP	ON	12.	NP/PP	POP		21.	NP/PP	POP	
5.	PP	CAT PREP	ON	13.	S/NP	CAT V	INDICATES	22.	S/VP	POP	
6.	PP/PREP	PUSH NP	DIODE	14.	S/V	PUSH NP	CATHODE				

Reasoning

Manufacturing knowledge is represented by networks and production rules. These networks and production rules provide the domain specific knowledge required to generate manufacturing plans automatically from the engineering database.

Networks describe the possible flow of parts through manufacturing and are composed of nodes which represent actions (operations) or decisions (movements). Arcs denote an association between two nodes. Networks can also depict the importance of precedent and sequel relationships. For example, a node placed in front of another implies a sequence of events.

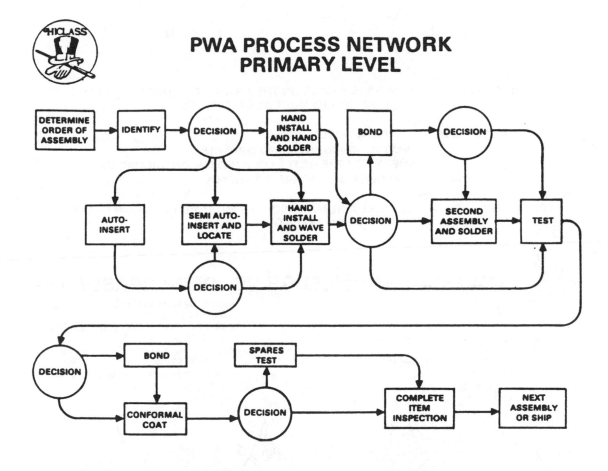

PWA PROCESS NETWORK PRIMARY LEVEL

PERT diagrams are a common example of a network and have been used for many years to analyze factory flow. Unlike PERT diagrams, these decision networks capture meta-knowledge to guide selection and evaluation of rules. The flow and control of a planner's reasoning can be modeled by such networks.

Networks can also be nested; that is, one network can be embedded within another. These structures provide multiple levels of abstration to a problem which relate to different levels of detail. At higher levels, for example, a network can represent the flow of parts through an entire factory. At lower levels, a network can represent a flow within a work center.

HOW DO WE MAKE THIS DECISION?

HUGHES

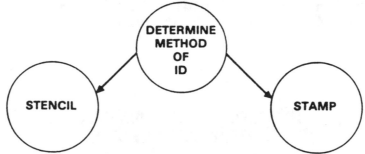

RULE OF THUMB: IF THERE IS A NOTE ON THE DRAWING "STENCIL BOARD . . . " THEN THE BOARD MUST BE STENCILLED

DEEPER QUESTIONS: WHAT IS A NOTE?
WHAT IS AN ENGINEERING DRAWING?
WHERE ARE THE NOTES LOCATED ON THE DRAWING?
HOW DOES ONE INTERPRET THEM?

IMPLICATION: JUST LIKE A HUMAN, HICLASS MUST BE TAUGHT IN ORDER TO BECOME AN EXPERT!

CAVEMAN KNOWLEDGE VS. DEEP REASONING

ASSERT HEADACHE COULD BE CAUSED BY PRESSURE . . .
ASSERT PRESSURE COULD BE CAUSED BY VESSELS DILATION
ASSERT ERGOT ALKALOID CAUSES VESSELS CONSTRICTION
ASSERT BARK CONTAINS ERGOT ALKALOID

IF I HAVE A HEADACHE THEN CHEW ON BARK

IF . . . THEN CHEW ON BARK

112

Manufacturing experience and intuition are fed to the inference engine in the form of production rules. IF-THEN logic is an effective and common method of representing this heuristic knowledge. In an expert system the IF statement is called the antecedent, while the THEN statement is the consequent. As with high-level programming statements, both the antecedent and the consequent can contain complex logical expressions. These functions manipulate and evaluate slots, which resemble variable names in high-level languages, however, this IF-THEN logic is evaluated in a declarative fashion, so ordering of the IF-THEN logic is unimportant.

Other functions further permit the software to mimic expert reasoning. Solutions to problems are rarely a straight "yes" or "no". Instead, humans arrive at conclusions based on partial or uncertain evidence. Fuzzy-set logic is designed to deal with problems where there are many shades of gray, such as in manufacturing planning. Production rules contain "certainty" and "threshhold" to facilitate fuzzy-set logic.

HICLASS INFERENCE ENGINE CAPABILITIES

- IF ANTECEDENT THEN CONSEQUENT

- COMPLEX LOGIC EXPRESSION

- FUZZY MODEL (CERTAINTY AND THRESHOLD)

- LIMITED LEARNING (EXPERIENCE TABLE)

- STABILITY ANALYSIS

- RESOLVE CONFLICTS (PRIORITIES)

- MULTIPLE LINES OF REASONING

THE KNOWLEDGE BASE

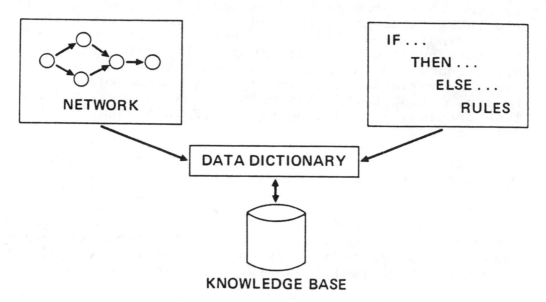

NETWORK

IF ...
 THEN ...
  ELSE ...
   RULES

DATA DICTIONARY

KNOWLEDGE BASE

EXAMPLE OF KNOWLEDGE STORAGE STRUCTURE

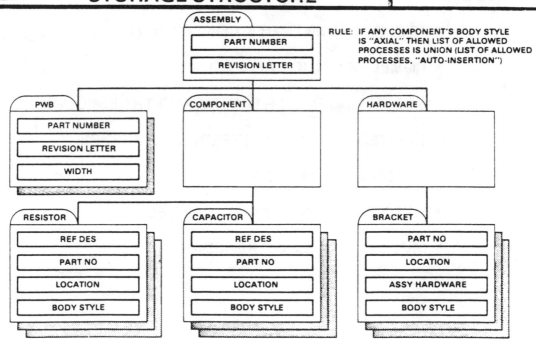

ASSEMBLY
- PART NUMBER
- REVISION LETTER

RULE: IF ANY COMPONENT'S BODY STYLE IS "AXIAL" THEN LIST OF ALLOWED PROCESSES IS UNION (LIST OF ALLOWED PROCESSES, "AUTO-INSERTION")

PWB
- PART NUMBER
- REVISION LETTER
- WIDTH

COMPONENT

HARDWARE

RESISTOR
- REF DES
- PART NO
- LOCATION
- BODY STYLE

CAPACITOR
- REF DES
- PART NO
- LOCATION
- BODY STYLE

BRACKET
- PART NO
- LOCATION
- ASSY HARDWARE
- BODY STYLE

Another problem the software must handle is non-determinism, which occurs when a single event results in a number of alternatives. For example, when the expert system recognizes that a cylindrical cavity needs to be created in a part, the alternative operations are drill, bore, broach, or ream. The system uses a ranking of priorities to resolve these conflicts. If priority is based on lowest cost alone, the system will choose drilling.

The expert system uses forward chaining (or forward inferencing) to generate manufacturing plans since the task involves creating a strategy for manufacturing a part from start to finish. In PCB assembly, for example, a planner starts with a bare board and works with it until all components have been assembled. Expert systems can also be built with backward inferencing, in which a problem is worked through from finish to start. In such a system, the PCB assembly would regress until it became a bare board. Since planners are unaccustomed to working with backward inferencing, it was not used.

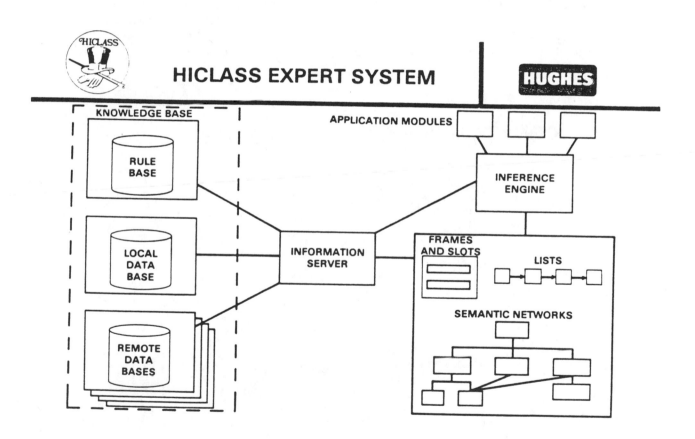

Man-Machine Interface

Automation of manufacturing planning involves computer interfaces on three levels. Experts (manufacturing and design engineers) must "teach" the system their knowledge. End users (workers who assemble the product) receive the finished plans. System implementers (typically computer scientists) develop the software which transforms expert knowledge into the finished plan. Since the computer literacy for each class of user varies, the interfaces much match the level of each.

The interface for experts is designed to acquire knowledge quickly. Typically, manufacturing engineers can best describe knowledge with PERT diagrams (process networks) and English-like statements for production rules. These experts use graphics-oriented engineering workstations to enter process networks into the system. One benefit of the system is that the knowledge representation is explicit and not implicit. In other words, these networks are used directly and do not have to be transformed into other knowledge representations.

Thus, no information is lost since there is no mapping involved. In addition, manufacturing engineers teach the system discrete logic required for manufacturing. This logic is represented by production rules.

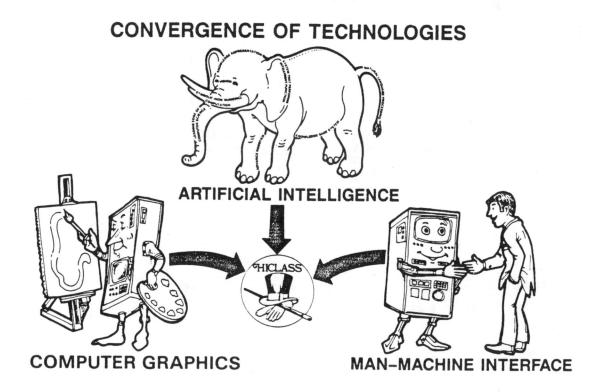

CONVERGENCE OF TECHNOLOGIES

ARTIFICIAL INTELLIGENCE

COMPUTER GRAPHICS

MAN-MACHINE INTERFACE

Presentation

The end-user interface is designed for teaching and presenting instructions on how to build a product. The product can be built by either humans or robots. For now, we will concentrate on the human aspects. The instructions can be relayed verbally, textually, or graphically. Hughes decided that pictorial representations are such the most efficient method of presenting information, so the expert system generates computer graphics illustrations along with textual instructions for production workers.

In addition to static frames of instructions, the system uses a script to embody information on dynamic rotation, translation, and other commands. Such functions require the host to have high-computational speed because software performs a large number of matrix manipulations to achieve graphical mapping. The hardware utilizes floating-point features and the operating system utilizes large virtual memory to enhance the functionality of the system.

High throughput is also important. The CRT at the factory workstation must display an image as fast as the host receives commands from the production operator. To accomplish this, these terminals must have graphics routines in both hardware and firmware. Since high computational speed and throughput are conflicting requirements in most computers, the systems uses a computer with powerful CPU's coupled to direct memory access (DMA) devices to meet the performance requirements.

Computer Graphics

Graphics are drawn on interactive systems amd are passed to the expert system in either a vector format or through the Initial Graphics Exchange Standard (IGES). For presentation, the system translates results from the decision-making process and combines then with results of the Rule-Driven Graphics Generator module to create assembly scenarios. The System now uses a Plot-10 graphics protocol to send these images to factory-floor terminals. In the future, the system may migrate to some video protocol such as the North American Presentation Level Protocol Syntax (NAPLPS) as well as video disk technology.

Conclusion

Expert systems are an excellent tool for automation. They embody human expertise to perform complex tasks such as manufacturing planning. The expert system holds great potential in the endeavor to automate manufacturing support services (white collar functions).

A REAL <u>EXPERT</u> SYSTEM

DISADVANTAGES

REQUIRES LOTS OF EFFORT TO TEACH THE SYSTEM THE BASICS

ADVANTAGES

HICLASS WILL NOT QUIT AFTER 2 YEARS OR RETIRE

UNIFORM APPLICATION OF EXPERTISE

CONSISTENTLY ACCURATE PERFORMANCE

ACCUMULATES KNOWLEDGE FROM MANY EXPERTS

INTEGRATES KNOWLEDGE FROM MANY DISCIPLINES

POTENTIAL AI APPLICATIONS

ENGINEERING
- CHOOSING SPECIFICATIONS, MATERIALS, PROCESSES
- DESIGN ANALYSIS
- PRODUCIBILITY ANALYSIS

PRODUCTION CONTROL
- PRODUCTION PLANNING

MANUFACTURING ENGINEERING
- CREATION OF PLANNING
- FINAL SELECTION OF MATERIALS, PROCESSES
- DESIGN OF TOOLING FIXTURES
- INTERPRETATION OF DESIGN PARAMETERS

ASSEMBLY AND FABRICATION
- SCHEDULING
- ALTERNATE ROUTING
- FAULT DIAGNOSIS
- ADAPTIVE CONTROL
- EQUIPMENT PROGRAMMING

QUALITY
- INSPECTION INSTRUCTIONS

Presented at the CASA/SME AUTOFACT '86 Conference, November 1986

Automating the Scheduling of Parallel Machines

by Mitchell S. Steffen
and
Timothy J. Greene
Carnegie Group Inc.

Computer-based scheduling systems frequently incorporate rules-of-thumb or heuristics to deal successfully with the complexity of manufacturing scheduling problems. This paper describes a heuristic that could be effective for scheduling many types of parallel machine systems. The heuristic is based on the observation that in some parallel machine systems certain machines tend to be temporarily dedicated to certain products during specific periods of time for quality or efficiency reasons. The subjective nature of defining which machines are dedicated suggests the use of artificial intelligence techniques. A case study, which illustrates an application of the heuristic, is discusses.

INTRODUCTION

In many industries, the complexity of production scheduling problems make them attractive targets for computerization. The introduction of MRP systems has automated much of the clerical work and routine decision making, but there has been limited success in automating the decision making required in the finite scheduling and dynamic realms [1, 5, 6]. The limited success is due to the complex and dynamic nature of many scheduling problems. The growing need for control flexibility in production systems is increasing the complexity of scheduling problems, creating an even wider gap between applications and the tools available for developing computer-based systems.

Developers of computer-based scheduling and sequencing systems typically use one or more of the following techniques: (i) mathematical models developed through scheduling theory, (ii) dispatching rules analyzed by discrete event simulation models, and (iii) human-computer interaction. While there are certain, very specific problems that can be solved by one or a combination of these techniques, many industrial scheduling problems cannot be effectively solved by these methods alone [1, 5, 6].

In manual scheduling systems, human schedulers attempt to deal with the problem complexity and volume of data by using rules-of-thumb (heuristics) that are based on successful past experiences. While some of these heuristics can become obsolete overnight due to product or process changes, others are more general and apply to a variety of situations. However, when development of a computer-based scheduling system begins, these heuristics are often overlooked and no attempt is made to evaluate their applicability in the new automated system. One possible reason for this situation is the difficulty of expressing and using subjective and sometimes imprecisely defined heuristics. The recent development of artificial intelligence-based tools and techniques can help change this situation.

This paper introduces a general heuristic for simplifying the structure of parallel machine scheduling problems, and discusses how to apply it in developing computer-based scheduling systems. The essence of the heuristic is to divide the scheduling problem into two parts: machines that are temporarily dedicated to specific products, and machines that make multiple product types. For most parallel machine scheduling problems, this "dedicated-shared decomposition" is not directly applicable unless supplemented with other techniques, because the resulting sub-problems many not be completely independent. In this paper, an artificial intelligence-based framework for implementing this heuristic is described. To illustrate the use of this heuristic, a prototype scheduling system for a case study is described.

PROBLEM STATEMENT

Parallel machine scheduling problems occur throughout industry. Research on scheduling methods for such parallel machine systems has traditionally assumed that any job can be made equally well on any machine in the system. However, in industry, systems of parallel but non-identical machines frequently exist [3, 8]. Further, systems that initially are "identical" tend to become "non-identical" over time due to machine wear and the addition of newer equipment. Good schedules for such systems must take into account the differences between machines.

To meet due date, quality or machine utilization requirements, human schedulers will often take advantage of differences between machines. For example, high priority jobs approaching their due dates would be assigned to faster machines, or jobs with tighter quality requirements would be assigned to machines that can hold closer tolerances. When viewed over time, such schedules appear to temporarily "dedicate" specific machines to specific products or job classes, while the remaining machines are "shared" among the remaining job classes. By dividing or "decomposing" the scheduling problem into dedicated machine and shared machine parts, large reductions in the effort required to solve the problem can be achieved. These reductions can improve the performance of a computer-based scheduling system for such problems.

While this dedicated-shared decomposition scheduling heuristic is conceptually simple, its implementation is not obvious for a given manufacturing system. The key to applying a dedicated-shared decomposition to a problem is the recognition that *specific machines are only dedicated for fixed periods of time*. As product demands and mixes change in the short term, and product and process designs change in the medium to long term, the current criteria for dedicating machines changes. The typical time span for the short term changes will probably be less than the planning horizon used in scheduling the manufacturing system. The challenge in using this heuristic in a computer-based scheduling system is to develop a framework that can reason with subsets of the planning horizon to allow the re-definition of what a dedicated machine is, and which machines should be dedicated.

APPROACH TO PROBLEM

Figure 1 depicts two typical hierarchies for solving industrial planning and scheduling problems. In this paper, production planning is defined to include those activities involved in master planning, capacity planning and inventory management. Loading is defined to mean the assignment of a job to a machine without regard to the timing or sequencing of the job. Sequencing is defined to mean the assignment of a specific ordering of jobs previously loaded on a machine, and if necessary, the determination of the job start times. Scheduling is defined to mean performing loading and sequencing simultaneously. The division of a scheduling problem into separate loading and sequencing problems implies that jobs can be assigned to machines without regard for their order or timing.

Figure 2 depicts two similiar hierarchies that incorporate a dedicated-shared decomposition. The advantage of adding an extra level to the hierarchy is that the resulting problems are each smaller than the original problem. Another advantage is that separate objectives and constraints can be used for the dedicated and shared problems, allowing a more focused approach to each problem. This heuristic could be viewed as a special case of resource-based decomposition [10] in that it divides the scheduling problem into smaller problems based on the machines, or resources, to be scheduled. The mathematics of complexity theory show that the effort required to solve two halves of a problem is less than the efforted required to solve the complete problem. In other words, "the sum is less than the whole".

As an example, a four machine problem with 12 jobs could have 479,001,600 possible unique schedules, assuming that each machine is assigned three jobs (real problems typically are much more complex than this example). If two machines are dedicated to six jobs, then the dedicated and shared scheduling problems would each have 720 possible unique schedules each, for a total of 1440 schedules to evaluate (plus the overhead from setting up the two problems). This is 0.0003 percent of the size of the original problem, or a five order of magnitude reduction in problem complexity, which corresponds to a dramatic savings in the effort or CPU time required to solve the problem. While this example over-simplifies several issues, it illustrates the potential benefits of using heuristics to decompose scheduling problems into smaller sub-problems.

The primary pitfall that can limit any heuristic's ability to decompose a scheduling problem is the occurance of interactions between the sub-problems. For example, if the queue order or the timing of a job has a major impact on which machine it is assigned to, then the scheduling problem cannot realistically be divided into separate loading and sequencing problems. When using a dedicated-shared decomposition, the system designers must be aware of any impact that the machines and products assigned to the dedicated machine scheduling problem have on making schedules for the remaining

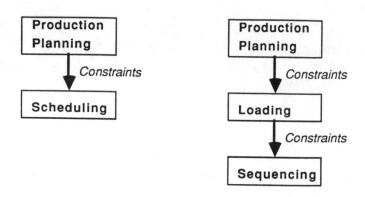

Figure 1: Common Planning and Scheduling Hierarchies

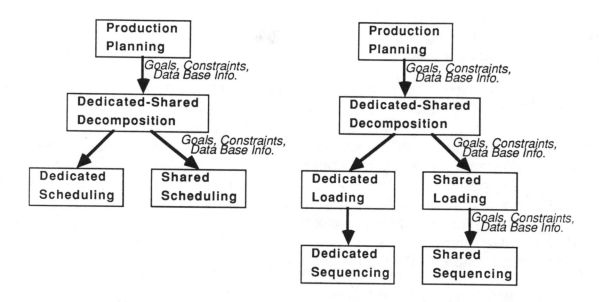

Figure 2: Hierarchies Incorporating a Dedicated-Shared Decomposition

machines and products. Typically, the problem characteristics which suggest using this heuristic also create a problem structure that has little interaction, or interactions that are resolvable.

A variety of situations could potentially benefit from the application of this heuristic. The applicability of the heuristic is indicated by the presence of a dominating process or product characteristic. Some examples of process-driven characteristics include:

1. Faster machines are often assigned to products with impending due dates.

2. The best or newest machines are often assigned to products requiring the highest quality.

3. The worst or oldest machines are often assigned to products with the lowest quality requirements.

4. Certain products require special operations that only certain machines can perform.

5. High profit products often are assigned to specific machines.

Some examples of product-driven criteria, which could arise in either identical or non-identical parallel machines systems include:

1. Schedule stability and/or consistency are important system objectives.

2. The application of Just-In-Time or Group Technology approaches requires specific machines to be assigned to "dedicated flowlines" for short periods of time.

3. Long set-up time products often are grouped on specific machines.

4. Specific products often are grouped on specific machines to achieve good utilization.

5. Products whose future demands are well known often are assigned to specific machines.

6. Products with unusually long or short cycle times often are grouped on specific machines.

7. Physical segregation of certain products is needed to maintain quality or minimize contamination.

Although the application of AI technology to actual industrial scheduling problems is in its infancy [11], we suggest using an artificial intelligence-based approach to implementing this heuristic for the following reasons:

1. The determination of which machines should be dedicated will probably be based on qualitative as well as quantitative criteria. AI techniques provide a variety of means to express subjective, non-quantitative and heuristic information.

2. The application of this heuristic will probably require repeated experiments with a prototype to find a workable design. AI-based programming environments provide powerful tools for rapid prototyping and testing alternative approaches.

3. Representing a scheduling problem's objectives and constraints in two separate problems (dedicated and shared) requires flexible representational methods. AI techniques provide a wide variety of means to represent and use constraints in scheduling problems [4].

4. If the dedicated machine and shared machine scheduling sub-problems are not completely independent (which is likely to be the case), they will need to be coordinated in some fashion. AI techniques provide a good framework for dividing a problem into sub-problems, switching between partially solved sub-problems and integrating the results [14].

Figure 3 outlines the functions needed in an AI-based approach to developing a decomposition module. The functions include: communication with the production planning system, defining which machines and products or jobs should be dedicated, assigning products or jobs to dedicated machines, and a knowledge base to support these functions.

Many companies already have some form of computer-based production planning, such as MRP II. Any scheduling system must interface with the existing production planning system in order to obtain information on orders, inventory, bills of material, master production schedules, etc. The communications function provides the link to the production planning data base. It also must translate the information into the format used by the knowledge base. Production planning data bases typically contain large amounts of very detailed information. The knowledge base may need some of this information in a summarized format. If so, the communications function would perform this service.

The knowledge base provides the information needed by the dedicate and assignment functions. The advantages of using a knowledge base over a standard data base include the greater ease of representing and maintaining qualitative and relational information and the ability to apply this information in problem solving. A variety of literature, products, classes and consulting services are currently available to assist in building knowledge-based systems and expert systems.

The criteria for choosing dedicated machines and the products or jobs to to be scheduled on them will most likely be empirically derived from observations of past system performance and human schedulers' experience. An expert system approach would probably be the most effective means to build an initial implementation of these criteria. Use of the expert system for a period of time may lead to insights into the criteria which could lead to the development of a more formal quantitatively based model. If this model proves tractible, and flexibe enough to keep up with changes in the system, then it should provide a faster alternative to the expert system.

The criteria for assigning products or jobs to dedicated machines may be closely related to the criteria for choosing them. In this case, assignment would take place in the expert system described above. Otherwise, the assignment would be cast as a "first-cut" loading problem in that the choosen products are assigned to dedicated processors without regard to time or queue order.

This section has identified the potential applications and benefits of using a dedicated-shared decomposition in scheduling systems for parallel machines, and outlined an AI-based structure for implementing this heuristic. Further expansion of the architecture of the decomposition module is application specific, and is best illustrated with a case study.

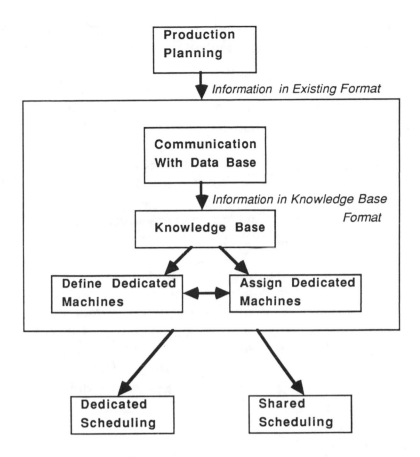

Figure 3: AI-Based Module For Dedicated-Shared Decomposition

<u>CASE STUDY</u>

The following sections outline an application of dedicated-shared decomposition to a case study problem to illustrate its use in simplifying the scheduling of parallel machines. The case study is described in more detail in [12]. It provides a good test of the heuristic because the human scheduler uses a dedicated-shared decomposition when manually creating the schedule. He starts by identifying which machines should be dedicated, then schedules these machines, and finishes by scheduling the remaining machines.

The case study production operation is a department within a large bulk material process operation. The facility operates on a continuous, 24-hour per day, seven days per week basis. The department converts a liquid raw material into powder form for down-stream departments, hereafter referred to as "consumers". The addition of special ingredients gives rise to a variety of final products. To meet consumer demand, 10 to 15 different products, out of about 150 possible products, are in production at any given time.

While the department has three stages of parallel machines, because of technological and organizational constraints they are treated from a scheduling viewpoint as a single set of parallel machines. Output buffers store completed batches. Each buffer supplies one consumer that draws the product from the buffers at a continuous rate. The consumers make no attempt to match this rate with the machines' batch size. Figure 4 is a schematic of the "logical" system.

In general, any machine can produce any of the products required by any of the consumers, but slight differences between machines result in the consumers preferring specific machines as their product source. Producing batches on non-preferred machines causes down-stream quality problems, so consistently satisfying these preferences is the human scheduler's primary objective. The set-up costs resulting from allowing an output buffer to be exhausted are prohibitive. Therefore, another important objective of the human scheduler is to maintain a sufficient safety stock in each buffer. The scheduler also considers several secondary objectives, such as minimizing material handling and minimizing the labor required to execute the schedule.

Several constraints limit achievement of these objectives. Incompatibilities between products prohibit certain sequences of products on the same machine. The output buffers have fixed, finite capacities. Power consumption and organizational needs require that machine batch cycle times begin and end at prescribed times as given in a standard schedule form, without preemption. Certain machines are not equipped to make products requiring extra ingredients. The output material handling system becomes congested when too many batches are scheduled in machines on the opposite side of the facility from the destination buffer. Exactly 100 percent of machine capacity must be scheduled.

In this case study, a dedicated machine is defined as a machine which is preferred by a consumer and the consumer's demand is greater than or equal to the machine's capacity during the planning period in question. A shared machine is anything leftover; that is, any machine that is not preferred by a consumer, or is preferred but the consumer's demand is less than the capacity of the machine. Quality constraints drive consumers preferences for specific machines. The non-identical nature of the machines results in slight product variations, which the consumers must compensate for to remain within quality specifications. By dedicating specific machines to specific consumers, little or no compensation by consumers is required, resulting in better quality final products and fewer rejects.

CASE STUDY PROTOTYPE DESIGN

Application of the dedicated-shared decomposition heuristic to this case study problem has progressed as far the development of an initial prototype for demonstration purposes. This section describes the design and development of the prototype scheduling system.

The prototype scheduling system was written in Virginia Tech Prolog [9], which runs in an interpreted mode on a VAX 11/780 under VMS. Several features of Prolog allowed rapid development of the prototype, including:

- Facility for expressing arbitrary relationships between objects as predicate clauses.

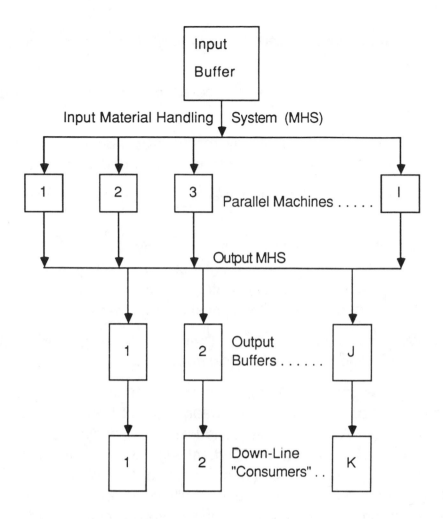

Figure 4: Schematic of Case Study From Scheduling Viewpoint

- Pattern matching capabilities.

- List processing functions.

- Automatic backtracking of program control when a line of execution fails.

The prototype makes extensive use of these features to develop and search the sets of feasible schedules. Clocksin and Mellish's text [2] provides a good introduction to Prolog. To supplement the facilities of Prolog, some relational table manipulation and execution tracing routines from the GUESS/1 [7] expert system development shell were used.

The heuristic was implemented using the following functions, executed in the order given:

1. Production Planning.

   In the case study, a computer-based production planning system did not exist. Information on product demand, department capacity and product inventory were manually gathered from a variety of sources and input into the prototype.

2. Identify which machines should be dedicated to one product type during the current planning horizon.

   The rule-of-thumb used was that any machine scheduled to be on-line throughout the planning horizon was a candidate for dedication.

3. Calculate the number of dedicated machines each product line would ideally need.

   For each product line, the demand is divided by the equvalent of the capacity of one machine on-line throughtout the planning horizon, and rounded up by adding 0.7 and truncating. This gives the number of machines that a product line could use most of the capacity of.

4. Assign dedicated machines to product lines in order of product priority, based on product preferences for specific machines.

   The purpose of assigning product lines to machines in priority order was to minimize the effects of conflicts between product lines that preferred the same machines. A better quality schedule for the dedicated machines could be obtained by using the Assignment Algorithm (a special case of linear programming), but this "priority order rule-of-thumb" required much less programming and testing, and gave results good enough to serve the prototype's purpose. Later versions of the prototype could be improved by adding the Assignment Algorithm.

   Because all jobs within a product line are identical, no sequencing is necessary. Therefore, the assignment of dedicated machines to product lines creates a complete schedule for these machines. In this case study, little effort is needed to schedule the dedicated machines in comparison to the shared machines.

5. Schedule the remaining machines and product lines.

   This requires two steps: assigning or loading individual jobs to machines, and then sequencing the jobs assigned to each machine. Constraint-directed search [4] was used to solve each of these steps. While the loading and sequencing problems are not independent, the prototype addressed the interactions by treating the loading assignments as "guidance constraints" [13] that could be relaxed during sequencing.

6. Schedule the miscellaneous jobs using the remaining machine capacity.

   Miscellaneous jobs are those not associated with a specific product line. Therefore, only a few scheduling constraints apply to them and they can safely be ignored until all other scheduling decisions have been made.

They provide an easy means to fill in holes left in the schedule by the previous functions (which in turn simplifies the logic needed by the previous functions because they do not need to make sure that all machine capacity is utilized).

CASE STUDY PROTOTYPE RESULTS

Evaluation of the prototype scheduling system was based on a comparison with schedules created by the human scheduler. Because the management of the production operation did not use quantitative measures of schedule quality, schedules created by the human scheduler provided the baseline of performance expected from the prototype. Tests of the prototype used eight days of actual data for which the human scheduler provided detailed notes on how he created the schedules. These eight days included many of the events that the scheduler normally must deal with, such as: changes in consumer demand type or quantity, machine status changes (on or off-line), requests for special batches, and changes in miscellaneous batch demand type or quantity.

Because the scheduling problem was modeled as a constraint satisfaction problem, adherence to the constraints included in the prototype formed the basis for comparison between the prototype and the human scheduler. Each schedule was analyzed to determine if constraints were adhered to, and if not, the degree to which the constraints were relaxed. In general, the prototype was able to find schedules that satisfied the problem constraints.

When the prototype needed to relax constraints, it did so to about the same degree as the human scheduler. Both the prototype and the human scheduler relaxed 0.1% of the instances of preference constraints, while the prototype relaxed 2.0% of the buffer safety stock constraints verses 1.3% for the human scheduler. The prototype has no logic for relaxing the buffer capacity constraint, and therefore created schedules that insured that batches were scheduled only when room was available for them in the buffers. However, the human scheduler took advantage of the fact that an implicit overflow capacity exists in the material handling system, allowing the build up of extra safety stock. For 11.5% of the batches scheduled by the human, the buffer capacity constraint was exceeded.

Considering that the prototype scheduling system didn't relax constraints significantly more than the human scheduler while it had one less degree of freedom than the human scheduler (it couldn't build up extra safety stock in the material handling system), the prototype's performance for the available test data demonstrates that a dedicated-shared decomposition approach has good potential for this case study problem. The results are discussed in more detail in [13].

CONCLUSIONS

Dedicated-shared decomposition is a technique currently used in manual scheduling systems as a way to reduce the effort required and to focus attention on the system's objectives and constraints. This paper has defined this technique and outlined an approach for applying it in the development of computer-based scheduling systems for parallel machines. Use of this technique can reduce the computer time required to create schedules by several fold. This improvement can help make real-time and interactive scheduling systems more feasible, or allow scheduling systems to evaluate more constraints to improve the quality of their schedules.

ACKNOWLEDGEMENTS

This research was funded in part by a Manufacturing Systems Engineering Intern Fellowship from E.I. DuPont de Nemours, Inc., in conjunction with the Department of Industrial Engineering and Operations Research at Virginia Tech. The authors thank Carnegie Group technical directors Ivan Johnson and Arvind Sathi for their contributions to this analysis.

BIBLIOGRAPHY

1. Browne, J., J. E. Boon, and B. J. Davies. "Job Shop Control". *International Journal of Production Research 19*, 6 (1981), 633-643.

2. Clocksin, W. F. and C. S. Mellish. *Programming in Prolog.* Springer-Verlag, New York, 1981.

3. Egbelu, P. J. "Planning for Machining in a Multijob, Multimachine Manufacturing Environment". *Journal of Manufacturing Systems 5*, 1 (1986), 1-13.

4. Fox, M. S. *Constraint Directed Search: A Case Study of Job-Shop Scheduling.* Ph.D. Th., Carnegie Mellon University, Pittsburgh, PA, October 1983.

5. Graves, S. C. "A Review of Production Scheduling". *Operations Research 29*, 4 (1981), 646-675.

6. King, J. R. "The Theory-Practice Gap in Job Shop Scheduling". *Production Engineering 5* (1976), 138-143.

7. Lee, N. S. and J. W. Roach. GUESS/1: A General Purpose Expert System Shell. TR-85-3, Dept. of Computer Science, Virginia Tech, 1985.

8. Reklaitis, G. V. "Review of Scheduling Process Operations". *Computer-Aided Process Design and Analysis 78*, 214 (1982), 119-133. AIChE Symposium Series.

9. Roach, J. W. and G. Fowler. HC: The Virginia Tech Prolog/Lisp Manual. Dept. of Computer Science, Virginia Tech, 1983.

10. Smith, S. F., P. S. Ow, C. LePape, B. McLaren and N. Muscettola. Integrating Multiple Scheduling Perspectives to Generate Detailed Production Plans. Artificial Intelligence in Manufacturing, Society of Manufacturing Engineers, Long Beach, CA, September, 1986.

11. Steffen, M. S. A Survey of Artificial Intelligence-Based Scheduling Systems. To appear in the Proceedings of the 1986 Fall Industrial Engineering Conference, Boston, MA, December, 1986.

12. Steffen, M. S. and T. J. Greene. A Prototype System for Scheduling Parallel Processors Using Artificial Intelligence Methods. 1986 Spring Industrial Engineering Conference, Institute of Industrial Engineers, Dallas, TX, May, 1986, pp. 425-433.

13. Steffen, M. S. and T. J. Greene. Hierarchies of Sub-Periods in Constraint-directed Scheduling. Symposium on Real-Time Optimization in Automated Manufacturing Facilities, National Bureau of Standards, Gaithersburg, MD, January, 1986.

14. Stefik, M. S. "Planning With Constraints (MOLGEN Part 1)". *Artificial Intelligence 16* (1981), 111-140.

CHAPTER 4

ROBOTICS AND VISION

Presented at the RI/SME Robots 8 Conference, June 1984

Expert Systems and Robotics

by Barry Irvin Soroka
University of Southern California

What is AI?

Artificial Intelligence (AI) is the attempt to make machines do things which, if done by people, would be said to require intelligence. Scientists have sought this goal since before the advent of computers. A thermostat exhibits primitive intelligence, but Americans have no awe for this device. Sorting a list of names or adding a column of figures is now considered a mechanical task, inefficiently done by people. AI is thus a moving target, because once a task, however complex, has been mechanized, it no longer seems to require intelligence. Intelligent behavior involves a certain mystery as to how the task is done.

The Not So Glorious History of Artificial Intelligence

In the 1950s and 60s, AI research achieved publicity with attempts at **playing chess** and **translating among languages**. Those on the ground floor of AI eagerly predicted that a machine would shortly be the chess champion of the world; this goal has not yet been achieved. An early AI program translated a sentence from English to Russian and back again. Its input was:

The spirit is willing but the flesh is weak.

The output was:

The vodka is strong but the meat is rotten.

As a result of such early failures, AI had prophets but not profits, and industrial funding sought other research outlets, such as VLSI. Only a few government agencies were willing to invest in AI during the 1970s.

Expert Systems is the come-back of AI in the 1980s. The media are giving space to Expert Systems despite the lack of pictures. And industry has discovered Expert Systems, as evidenced by many start-up firms and industrial research groups in this specialty. The start-ups are betting that Expert Systems will peak in about 1986, predicting a boom rather than a bust. This paper is an introduction to this emerging field, with pointers for further investigation.

Why Expert Systems Work

Many problems involving intelligence can be seen as examples of *search*:

1. To find the best next move in chess, evaluate the consequences of all possible moves.

2. To determine a patient's disease, check the symptoms against all possible single illnesses and combinations.

3. To find the best layout of equipment for a factory, try all possible positionings and pick the best.

Over the years, AI has produced many algorithms for searching solution spaces of the type listed above. But for real problems, the solution spaces are too large to be handled by current or even future computer systems. Blind search is insufficient.

Humans do *not* exhaustively enumerate the possibilities when confronted with problems like these. Physicians use the patient's history and symptoms to narrow the choice of diagnosis. Applications engineers use experience to suggest good layouts of equipment on the factory floor. Chess grandmasters have an intuitive feel for what sort of move is appropriate. People are satisfied with sub-optimal solutions as long as those solutions are pretty good.

Expert Systems are an attempt to capture in programs the *heuristics* ("rules of thumb") used by experts in solving problems for which exhaustive search is impossible. This domain-specific knowledge is used to guide search through the otherwise intractable solution space. *Knowledge engineering* refers to the process of transferring an expert's knowledge from the head of the expert to the rules of an Expert System; it is not an easy task.

Current Expert Systems

The first Expert System, DENDRAL, was developed at Stanford to determine the structure of unknown chemical compounds on the basis of mass spectrographic data. This program became quite expert at its task, and the project is now sixteen years old.

Here are some current Expert Systems which have found commercial utility:

- XCON is an eXpert CONfigurer of VAX computers for DEC. *Each VAX is unique*, depending on the options selected by the customer. XCON takes a customer's order and selects the boards, slots, and cables required to produce a working system which meets specifications. Not all arrangements of boards and slots will work: if the cable between two boards is too long, then the system may not work when run at its expected clock rate.

- PROSPECTOR predicts the locations of ore deposits based on geological data from the site. One hears various accounts of how valuable are its finds.

- DELTA/CATS-1 is used by General Electric to diagnose faults in diesel locomotives. It prompts the user to supply instrument values and test conditions on the metal patient, then suggests fixes and monitors their success. An attached interactive videodisk system supplies still pictures, diagrams, and recorded demonstrations to help even a naive user perform the required diagnostic and correcting actions.

In general, Expert Systems are useful in the following sorts of tasks:

- debugging
- design
- diagnosis
- instruction
- interpretation
- monitoring
- prediction
- planning
- repair

Few types of intelligent activities seem to be missing from this list.

Inside an Expert System

Expert Systems consist of three major components:

- a database;

- rules for transforming the database;

- a control strategy for selecting which rule to apply next.

Individual Expert Systems differ along all three of these dimensions.

The **database** is a representation of parameters important to the problem. For example, the database of an Expert System for diagnosing problems with automobiles would contain facts such as these:

- The car starts.

- The car has gas.

- The battery is OK.

The **rules** of an Expert System encode knowledge about how to reason from one set of facts to another. In the auto example, typical rules might be:

- If the car won't start, then see if the car has gas.

- If the car won't start and the car has gas, then see if the battery is OK.

- If the battery is not OK, then replace the battery.

Notice that each rule represents a single piece of knowledge about the domain involved. Thus, *we can add knowledge incrementally by adding more rules.* This is very different from conventional computer programming, where deleting a subroutine typically results in a total loss, rather than a simple degrading, of performance.

The *control strategy* determines which rules to apply when several rules are applicable.

For example, physicians don't send a patient for a CAT scan merely because the patient has a headache. Nor do we presume that a patient has dengue fever merely because the temperature is elevated. Cheap tests are done before expensive ones, and common diseases are hypothesized before exotic ones.

Expert Systems in Robotics

Here are a few examples of robotics problems which may be helped by use of Expert Systems.

Kinematics and Design

A basic problem of robot kinematics is going from a desired cartesian position and orientation back to the joint angles required to achieve it. Straight-line motion requires knowledge of this function, the *arm solution*. For a given robot arm, the forward function which takes joint angles to end-effector position is nonlinear and transcendental; its inversion is analytically possible only for a small class of robot arms. Not surprisingly, only this class of robot arms is manufactured. Myths abound: for a long time it was believed that an arm could be solved only if its final three joints (wrist) intersected at a point; last year Intelledex laid that myth to rest.

Each kinematically new robot requires its own unique arm solution. Producing such solutions is symbolically tedious work, prone to error. Nonetheless, existing arm solutions have been obtained using only a few simple heuristics for searching a set of equations. An Expert System can probably be implemented to replace the human expert at this difficult task. Such a program could also be designed to explore the space of possible robot configurations, looking for designs which satisfied input criteria such as workspace shape and size.

Robot Selection

Many robot applications involve replacing a human operator with a robot. This happens, for example, when heavy machines cannot be rearranged. Rather than designing an ideal robot workplace from scratch, we continue the current arrangement and find a robot which fits. An Expert System can aid this process.

The program would prompt the user for important parameters of the required robot:

- reach;
- speed;
- number of program steps;
- payload capacity;
- repeatability;
- accuracy.

Additional constraints could also be described, such as:

- maximum space available;
- clean-room considerations.

Trade-offs exist concerning:

- new wiring or piping;
- special construction and installation;
- safety devices.

When selecting a robot, the human expert --- the applications engineer --- considers all these factors in determining cost, payback, and return on investment.

Experienced applications engineers are in short supply today. An Expert System could be useful in diffusing applications experience among a wider community of robot users.

Workplace Layout

Sometimes we design the robot workplace *de novo*, either for a new factory or a new line. There are fewer constraints on what robot we can use, because we can build around absent robot capabilities. For example, we can use a robot with lower repeatability if we spend more money on the fixtures with which it interacts.

There are also few constraints on how machines should be arranged; placement of machines has terrific impact on throughput and work-in-progress. In an extreme case, notice how the machine shop based on Group Technology principles has a crazier layout but greater throughput than the shop based on functional similarity of machines. Expert Systems will eventually be able to discover major manufacturing principles such as Group Technology on their own.

Clearly, applications engineers do not try all possible machine placements in laying out robot lines. There is little science for this task, but presumably planners use some rules of thumb which can be translated into heuristics for an Expert System to guide the search for useful factory layouts. The challenge is in the Knowledge Engineering: trying to make manifest the expertise used by expert application engineers.

Maintenance

Expert Systems have an obvious role to play in diagnosing robot malfunctions. Prepared by the manufacturer, such programs could lead the user through the symptoms of trouble, request test measurements, suggest adjustments and repairs, and monitor the process of fixing the robot. Interactive videodisk "manuals" and "blueprints" could be used to show maintenance what to expect during each step of disassembly and repair.

Many problems are presented by systems where a robot interacts with intelligent peripheral machinery. Timing bugs are extremely difficult to locate and fix. By the year 2010, most such systems might be adequately modelled on computers, and Expert Systems will incorporate more knowledge about concurrent processes and bus protocols than the human maintainers will ever possess. People will serve as the legs and hands of an overseeing Expert System brain which monitors the factory, finds problems, and determines their solution.

Robotics and AI

Some years ago, the editors of the *Artificial Intelligence Handbook* considered including a chapter on Robotics. The final version of the *Handbook* lacks such a chapter, and for good reason: the overlap of AI and Robotics has been practically NIL.

Historically, AI has been concerned with *symbols*, Robotics with *mathematical details*. AI works on *big machines* and languages like *LISP*, while Robotics relies on *small machines* and *assembly language* for speed. AI works on *high level* problems such as planning, while Robotics looks at *low level* problems such as friction and compliance. It is no wonder that the Robotics community has benefitted so little from AI research.

Problems with Expert Systems

The great bottleneck with implementing Expert Systems is Knowledge Engineering. There are too few people who know much about Expert Systems, how to write them, use them, or translate knowledge from the mind of the expert to the rules of an Expert System. Job competition is fierce: a student with a Master's degree and experience in Expert Systems receives terrific job offers.

Expert Systems are different from conventional computer programming. Heuristic rules are not the same as procedures and algorithms. Not all current programmers can make the transition to writing in Expert System style.

Will Expert Systems Make It?

Will AI be vindicated at last? Will a hundred million dollars of Federal funding finally find commercial applicability? Or will Expert Systems be just another AI fad? Here are a few reasons why AI *might* make it on this third attempt:

- We know much more about how to program intelligent behavior than we knew in the 50s and 60s. AI has learned from its mistakes.

- Computation is much cheaper today than it was even a decade ago.

- Computers are larger than ever. It is not inconceivable to put a miniVAX on a robot. LISP and other AI languages will soon be running on cheap computers. Programmers are not forced to use the low-level algorithmic languages of the 50s.

- Industry is more realistic in its expectations from new technology. We know there are no panaceas for manufacturing problems. The systems approach encourages us to look for significant incremental improvement where we can. No single technological thrust will solve all our problems, but used judiciously, Expert Systems can save us money.

Information Sources

Books

Books on Expert Systems are just beginning to appear. Here are two which are essential:

1. Frederick Hayes-Roth, Donald A Waterman, & Douglas B Lenat, *eds.*, *Building Expert Systems*. Reading MA: Addison-Wesley, 1983. Introduction to Expert Systems, building them, evaluating them, and a case study implemented in eight different Expert Systems. A Bible for Expert Systems.

2. William B Gevarter, *An Overview of Expert Systems*. NASA-NBS Report NBSIR 82-2505, May 1982. Probably available from NTIS. A concise summary of issues and existing Expert Systems (rapidly becoming dated).

Magazines and Journals

1. *AI Magazine*. Published by the American Association for Artificial Intelligence (see below). News and technical articles.

2. *SIGART News*. Published by ACM's SIGART (see below). Mostly abstracts of just-published technical papers.

3. *Artificial Intelligence Journal*. Occasional scholarly reports on Expert Systems work.

4. *Journal of Expert Systems*. Not yet announced, but no doubt coming soon.

Organizations

1. American Association for Artificial Intelligence (AAAI). 445 Burgess Drive, Menlo Park CA 94025. Founded about four years ago, they sponsor a conference and publish *AI Magazine*. Their mailing list finds its way into the hands of most of the AI consulting firms, vendors, and publishers.

2. Special Interest Group on Artificial Intelligence (SIGART) of the Association for Computing Machinery (ACM). 11 W 42nd St, New York NY 10036. They have chosen to shun the flashy route of expositions and conferences, targeting SIGART to the academic community. Considering the spectacular financial success of ACM's SIGGRAPH, this conservatism may not last long.

Conferences

1. IJCAI. International Joint Conference on Artificial Intelligence. Held every two years, its location cycling among North America, Europe, and Japan.

2. NCAI. National Conference on Artificial Intelligence. Sponsored by the AAAI. Held in the US in years when IJCAI is elsewhere. Papers plus exhibits. Last year's conference was in Washington DC. Industrial attendees outnumbered academics for the first time.

The proceedings of these conferences can be obtained in the US from William Kaufmann Inc, 95 First St, Los Altos CA 94022.

Technical Reports

Many advances in Expert Systems and AI in general are first seen in technical reports emanating from:

- Carnegie-Mellon University
- Massachusetts Institute of Technology
- Stanford University
- SRI International

Reprinted from the February 1985 issue of *Test & Measurement World*
Copyright 1985 by Cahners Publishing, a division of Reed Publishing USA

INSPECTION

Intelligent Vision Systems: Today and Tomorrow

Detroit's sudden demand for vision products took the machine vision industry by surprise. At that time, machine vision companies were just starting to use binary threshold, SRI algorithms and heavy up-front application engineering efforts. Spurred by the new opportunity, some 200 to 300 vision companies formed, 53 of which can now be considered as machine vision houses.

Lawrence A. Murray and James E. Cooper
Vuebotics Corp.

Early machine vision companies concentrated on specific functions rather than on general forms, developing many different machine vision units. Each unit was based on a single, specific, simple vision technique, such as Fourier transforms, histograms, templates and fixed photo-detectors focused on fixed positions. Later machine vision systems relied more heavily on pre-processing, feature extraction and development of user-friendly menus. In addition, feature extraction hardware became more independent and modular and began using dedicated bus structure architectures. Because most systems sold to the auto industry had rigidly fixtured part presentations, they used few image manipulation operations, and pre-placed windows played a significant role in parts analysis. Binary thresholding and two-dimensional views became the norm.

As machine vision system architecture became better defined, the need for a transition from machine vision to "intelligent vision" also became clear.

Moving from simple machine vision to intelligent vision requires:

Image acquisition: color

Pre-processing: gray scale, hardware connectivity, texture analysis

Feature extraction: three-dimensional (distance) measurement, touching-object recognition, non-rigid object recognition

Image operations: rotation, correlation, left/right reversal, computer-aided design/computer-aided manufacturing (CAD/CAM) interfaces, cellular automata

Image understanding: flexible algorithms, built-in-application-independent-programming, implicit inspection.

This list is not a complete recipe for intelligent vision. Some companies have addressed some of these needs and incorporated the solutions into

The joining of artificial intelligence algorithms and image processing techniques with charge coupled imaging devices (CCD) led to the birth of a new industry — image analysis.

their machine vision systems. The list points out the deficiencies of present-day intelligent vision models.

Historical Development

The joining of artificial intelligence algorithms and image processing techniques with charge coupled imaging devices (CCD) led to the birth of a new industry — image analysis. At the outset, the industry called itself "computer vision" because rapidly processing the complex algorithms required a powerful computer. Later, as the use of vision systems penetrated automotive production lines, people began to call the field "machine vision." Units that use only one analytic technique (such as windows, templates, histograms or optical couplers) became generally known as "simple machine vision" systems.

As vision systems continue to grow smarter and more flexible, to serve broader and more diverse industrial applications, the field is evolving into "intelligent vision." Intelligent vision systems serve such diverse industries as automotive assembly, personal computers, food processing, medical components, electronics and robotics. These systems can perform five independent functions: image acquisition, image pre-processing, feature extraction, image operation and image understanding.

Solid-state cameras, the key to image acquisition, come in a number of forms: charge coupled devices (CCD), charge injection devices (CID) and photodiode arrays (SSPD/MOS). Regardless of which of these technologies the camera uses, the camera lens focuses light on a chip divided into light-sensitive cells (pixels) causing an electrical charge proportional to the amount of light falling on each pixel to build up on it. In many cases, the number of pixels on a chip is too small to provide

Table 1: Applications Categories

QUESTION	APPLICATION	TYPICAL INDUSTRY
What (Who)	Identification	Assembly
Where	Location	Robotics
Which	Recognition	Sorting
When	Motion	Surveillance
How Good	Inspection	All Industries Except Robotics
How Large	Gauging	Industrial Control
How Many	Counting	Inventory

adequate image resolution. In such cases the image spaces of several solid-state cameras can electronically join together into a super-pixel space to resolve as large an object as required.

Image processors take a snapshot of a scene and process it electronically. Units are available which remove noise, sharpen edges, heighten contrasts, readjust or change colors and extract gray levels, store a scene and subtract it from or add it to the next scene, or compress data using techniques like binary thresholding with run-length encoding. The function of any image processor is simply to make the picture more readable. In a machine vision system, this image processing function precedes analysis and prepares the image to facilitate the main task of a vision system — image analysis.

The aim of machine vision is to recognize, gauge, locate, identify, inspect and count the objects in a scene. Connectivity defines a group of objects in the field of view so that the vision system can extract their pertinent features.

Artificial intelligence applied to machine vision includes automatic feature extraction, image operations, and image understanding. Image understanding is the frontier for vision research and future breakthroughs.

Applications

Table 1 shows how machine vision applications fall into different categories depending on what vision-related questions need to be answered. Other applications require solutions to more than one problem. For example, sorting consists of a mixture of location (where) and identification (what) or recognition (which), inventory consists of a mix of counting (how many) and identification (what). The questions "what" and "which" and their associated application areas

do not lend themselves to as sharp a distinction as the others do. "What" connotes general information about kind rather than degree (man, chair, auto, building), while "which" tends toward the specific and the degree-oriented (Adonis, cane-backed, Model T, Empire State building).

Markets

Target markets for the introduction of intelligent vision are industries that are especially concerned with productivity, quality and control.

In its intense struggle to wrest market share back from the Japanese, the American automobile industry decided to improve its quality through 100% inspection of all parts and assemblies. In addition, it has put pressure on its suppliers to provide high-quality raw materials, stampings and parts on time. This effort is providing hundreds of thousands of potential sites for machine vision systems. Two characteristics of the automotive industry had a profound influence on the early machine vision industry: its cycle times are slow (14 seconds between parts and 90 seconds between autos), and turnkey systems with highly fixtured presentations are the norm. The slow cycle times allow vision systems to do without high-speed processing. With speed less important, vision systems can have a high software content. Oriented presentation of parts and custom-fixtured illumination encourage heavy reliance on windowing and metrology rather than pre-processing or image operations. Furthermore, the turnkey mandate includes control of upstream and downstream equipment using interface software.

In the near future, the auto industry will probably place less emphasis on weeding out failures and more on building quality into its products. Machine vision will need to change

from parts inspection to on-line gauging of the tools and fixtures responsible for forming the parts — a move toward "in-process control."

The leading machine vision companies tend to be turnkey systems producers. The auto industry recently took percentage ownership positions in a number of such companies, and has special relationships with others.

The largest user of vision systems is the electronics industry. The micro-electronics capital equipment sector is particularly important. Because of the sophistication of this industry and its enormous throughput, it takes a very different approach to vision than the auto industry does. It tends toward OEM rather than end-user purchases. Electronics companies can add vision to their existing equipment rather than buying entire vision modules. They can also buy identical vision subsystems in quantity, rather than single custom systems. They look for prices in the range of $5000, rather than $150,000 per unit.

Vision for the electronics industry must take a snapshot, analyze it and make the location or GO/NOGO decision in milliseconds. The wire bonders, die pickers and sorters, pattern inspectors, automatic wire bond pullers and other kinds of equipment that permeate semiconductor manufacturing plants are essentially robots with eyes. The vision portion of such machinery contains a few circuit boards with statistical (pattern recognition) algorithms, rather than the stochastic (object recognition) algorithms common in other industries.

Electronics manufacturers began adding vision to their wire bonders over ten years ago, before the machine vision era. Systems only needed to recognize and orient a simple pattern for the first bondpad. After that bond was made, the internal microprocessor took control and reoriented for the prescribed bonding pattern. Since that time, great strides have been made in vision technology, and the equipment serving this industry should be upgraded to take advantage of those advances. Pressure from VHSIC (Very High Speed Integrated Circuit) programs that require careful inspection of all parts and from Mil-Spec requirements will ensure that this transition takes place.

Delphi surveys indicated that the future growth of the robot industry may depend on vision and that, by decade's end, half of the new robots will have sight.

Today, robots are generally large, heavy, multi-articulated arms with a feedback sensor in each joint. They have slow cycle times (4-6 sec) so attendant vision systems need not be fast. Since the robot has to move to where a part is, pick it up and take it to its next position, the vision system need only determine the part's location and orientation.

The interface between the vision unit and the robot represents an important component of any machine vision system. The two pieces must speak the same language. Most robot companies have developed their own languages, following no standard, and hence the leading robot houses try to develop their own vision systems and interfaces to fit their particular robot systems.

Although the popular press and the public in general tend to tie robots and vision together, there are and will be far fewer robots in operation than vision systems. The market for robots will grow more slowly than that for vision.

If a substantial market for sighted robots does arise, it will be for simple, light, fast, pick-and-place robots, such as those needed in the electronics industry. Vision will provide such robots with the intelligence to upgrade their operating skills and may, in some cases, obviate the need for mechanical stop mechanisms.

Developing Markets

For inspection, the parts made from CAD/CAM generated drawings are compared to a perfect, little-handled and carefully stored, "golden standard" or "golden image" of the part. Logically the drawing database can go directly into the vision module which can then inspect the finished part by comparison. Early applications involve two-dimensional parts, such as VHSIC reticles, printed circuit board patterns and hybrid circuits. Later applications will include three-dimensional parts, with the database rotated to align it to the part.

Computer memories can store databases in various formats — as simple centroids and distances, or as corners and distances to more complex structures. These formats require translation into some common format acceptable to vision, such as a pixel-based or a centroid-based format.

For example Optical Character Recognition (OCR) applications fall into one of two extreme categories.

> **The interface between the vision unit and the robot represents an important component. The two pieces must speak the same language.**

The characters are either high contrast, perfectly formed and well spaced (as in labels and printed pages) or poor contrast, ill-formed and smudged (as in hand letters and serial number stampings). Blob and feature analysis techniques suffice for reading the first kind of characters. The second kind requires careful pre-processing and custom algorithms that depend on the type face. Between those extremes lie such applications as reading laser-scribed serial numbers on silicon IC wafers.

OCR techniques appropriate for the first category seem straightforward compared to the algorithmic attack used for the second. Each character, whether uppercase, lowercase or script, has a set of uniquely describable features. The first and most obvious feature is the number of enclosed holes in the character — called "lakes" or "ponds." For instance, the characters A, B and C have 1, 2 and 0 lakes. Connectivity analysis automatically detects lakes.

The second most obvious feature is the number of "bays" or non-closed "inlets," also called "convex deficiencies" in convex hull analysis. For example, E has two bays pointing west; F has one west bay; and X has a bay in each cardinal direction. A small change in the connectivity algorithm allows detection of bays.

Developing Techniques

Such diverse applications as accurate inspection of automotive parts and recognition of smudged characters

require gray scaling. Some systems simply take account of incompletely filled edge pixels to increase feature (e.g., centroid, area) measurement accuracy. For example, PCB inspectors must distinguish the various parts of a completed board (light tan board, silver conductors, components of various shapes, sizes and colors, solder masks and solder) from one another.

Today, no major machine vision system includes color analysis because color CCD cameras have historically been too expensive and because color has found no place in binary threshold machine vision. In inspection applications, vision systems have checked parts with uniform (except for stains) metallic color. As vision moves into new industries, color will become an important differentiating characteristic.

Color cameras have three chromatic outputs that can be visualized on a three-axis color diagram. The ratios of the three outputs define the color, both mathematically and visually. Analysis can regionalize, separate or define these ratios using connectivity. The system can treat each region in the color diagram as an object and analyze it like any other object.

Since today's production lines emphasize object analysis and not scene analysis, machine vision seldom uses texture analysis. As vision begins to move out beyond its present limitations, the extraction of objects, patterns, foreground and background from a scene will become more important. Fourier analysis and histograms can both generate local signatures to identify textures. Connectivity analysis can discern these regions as objects or groups of objects.

Just a few years ago, potential customers commonly asked machine vision companies, "Is your system capable of working with six or more different objects in a single field of view, some overlapped and some touching?" Since then, hardware connectivity has demonstrated an ability to separate hundreds of objects so other imaging algorithms can analyze them individually. In most vision systems, two touching objects of the same thickness look like a single object that should be rejected. Gray scaling and cellular automata can sometimes separate them in particular applications. Rotating templates with modified configurational tolerances represent another interim solution. If the inner template fits the part (black region) and the outer part (white region) fits the template on three sides but is significantly out of tolerance on the fourth side, this side defines a line of osculation to separate the two objects. Any part under inspection is unlikely to be grossly out of tolerance in area, perimeter, radii, centroids and moments, simultaneously. The problem calls for an independent algorithm that can locate the line or points of osculation and redefine the combination as separate objects. The system could then determine the existence and location of the touching parts.

As machine vision sales escalate and modular pre-processing and feature extraction become standard, miniaturization of vision systems will become inevitable. Once volumes are high enough to write off the setup cost, a single VLSI gate array can replace 100 integrated circuits, 100 IC sockets, a printed circuit board an edge connector and all the labor involved in assembling them. The gate array approach also makes reverse engineering more difficult.

One section of the intelligent vision system that will resist miniaturization is "image understanding." This task does not lend itself to partitioning as

---

Any part under inspection is unlikely to be grossly out of tolerance in area, perimeter, radii, centroids and moments simultaneously.

---

neatly and successfully as applications, feature extractions, image operations, image acquisition and pre-processing.

Thus a vision system might consist of a tiny camera (with a big lens) containing a few chips that perform most of the data extraction, and a large vision module that receives the data and understands it. There are now VLSI chips for edge detection, video formatting and timing/driving image arrays. Both France and Japan have described the development of one or two chips for simple machine vision. Whether such custom chips will go into the camera, into the vision module or into their own module depends on the individual application.

Future Directions

Machine vision is moving toward intelligent vision, as shown in Figure

1. In the process, it is spinning off areas of vision to robots, cameras and turnkey systems houses.

Systems that emphasize pure location are already concentrated in the robotics industry. Systems houses will eschew R&D vision programs and will emphasize interfacing modules instead, selecting the simplest available. Camera companies will add pre-processing modules, extra modules and some simple image operation modules to the camera itself. This relatively long-range development presupposes a set of high-volume, low-cost applications that call for off-the-shelf cameras with little image understanding capacity.

Meanwhile, research on image understanding will seek to add to and upgrade vision intelligence. The goal of such research includes systems that can envision a scene, separate out its important objects, analyze them, understand them and then make the correct decision, with minimal programming. Such systems will not be ready in this decade, but, by 1990, some identifiably intelligent vision systems will be readily available.

Artificial Intelligence. Image understanding represents the most fertile ground for intelligent vision research. Many large universities, such as Rhode Island University, New York University, Carnegie-Mellon, MIT,

---

FIGURE 1. Evolution of the vision industry.

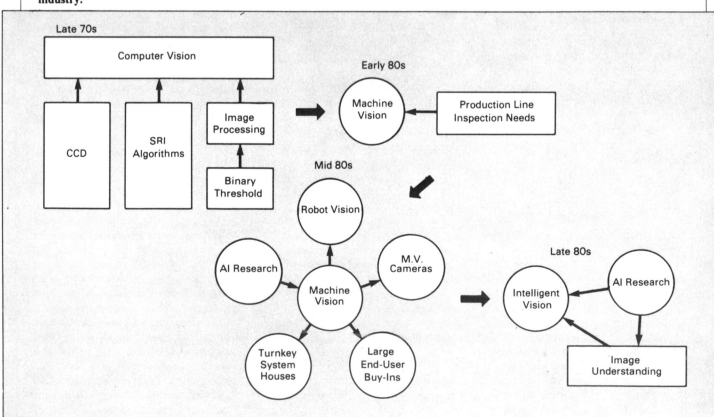

Stanford, the National Bureau of Standards and Stanford Research Institute (SRI) have established long-term programs in artificial intelligence (AI). These programs involve complex algorithms and rely on high powered computers. Naturally, they emphasize general AI, rather than vision. Some support comes from the government (e.g., DARPA and Night Vision Labs) which has targeted autonomous vehicles and three-dimensionality as early showpieces for this technology, as well as continuing efforts in constructing advanced expert systems.

AI often tries to generate programs analogous to human brain functions. Some are even neuro-physiologically based. But machine vision applications and processes have little in common with human-based perceptions, logic and decisions. It remains to be seen how compatible such AI research can be with intelligent vision system operation and goals. AI does not place much emphasis on solving vision-type problems because it regards vision more as a sensor than as a driving force.

Three-Dimensional Research. Almost all machine vision systems study two dimensional objects or the

Gray scale analysis gives three-dimensional information using the inverse square law of light intensity versus distance.

two-dimensional projections of three-dimensional objects. The one exception employs structured light patterns to infer distance information. Such systems typically find use in inspection.

Structured light, as shown in Figure 2, aims a line of laser light obliquely at a part as it moves down a conveyer belt. The height of the part offsets the light line in the camera image plane. The distance between the light on top of the object and the reference line indicates the height of the part.

Structured light has two important properties: It brings high contrast to the edges, and measures distances along the line-of-sight. The laser line is bright at the edge of the part, but dark

a short distance from the edge because of translation of the laser line. This change provides the vision system with a high contrast binary threshold outline of the part. The distance between the laser reflection position and the reference line for each point measures the height of each point. This is a specific, application oriented technique, even in the case of complex systems that use grids, arrays of points and multiple laser lines.

Stereoscopic vision seems like the most obvious answer to three-dimensionality because humans all use it. Unfortunately, its perceptions are subject to illusions, occlusions and shadowing misinformation. Stereo-vision requires heavy and slow computational analyses. Not troubled by human limitations, machine-oriented techniques or sensors could provide depth information more efficiently than techniques based on imitation of human systems.

Other lines of investigation include adaptive optics, dynamic focussing, gray scale analysis, and time of flight analysis.

Dynamic focusing comes in a variety of forms. One technique focuses the light from the target onto two sets of photodiodes. This system moves the optics gradually until the difference in output from the two sets reaches a minimum. This focal point can be converted to distance. This technique proves accurate for focusing onto a flat field, but can be adapted to moving the focus incrementally and snapshotting the focused section of the image.

Another method, adaptive optics, uses a lens with an extremely shallow depth of field. This system incrementally steps the focus from near to far field, and takes a snapshot at each step. The narrow focus washes out the foreground and background, and the focused region stands out in high contrast. These snapshots can be sent to the vision system for electronic reconstruction, since each snapshot is tagged with a distance value.

Gray scale analysis gives three-dimensional information using the inverse square law of light intensity versus distance. It is limited by its assumption of uniform shading and coloring, uniform reflectivity, no stray reflections or sharp corners.

Time-of-flight techniques use an auxiliary sensor and a laser. The system aims the laser at the target and step-scans the laser across the target horizontally and vertically, one line of

FIGURE 2. A structured light system.

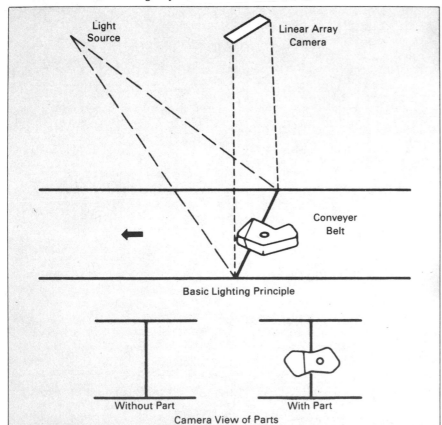

pixels at a time. The time of flight from the laser to the target and back to the sensor provides a direct measure of the distance to the target point. The rest of the vision system determines the x-y position of these pixels. The current time resolution of about one nanosecond corresponds to 30 cm. A vision system using phase modulation and a photodiode to detect the phase to an accuracy of one part in 1000 would give a distance resolution of 0.03 cm.

This article consists of edited excerpts from a book entitled "Intelligent Vision Systems for Industry" being prepared by Lawrence Murray for publication by Marcel Dekker in 1985.

LAWRENCE A. MURRAY *is Executive Vice President and Director of Advanced Technology of Vuebotics Corp., a company that manufactures machine vision systems. Prior to co-founding this company in 1981, he served as an electro-optic consultant for five years. Before that, he worked as General Manager of Vactec Advanced Technology Div. (St. Louis) for five years and was involved in micro-electronic and crystal growth research at RCA and ITT Labs for 16 years. He has written 60 technical papers, articles and presentations.*

JAMES E. COOPER, *Senior Scientist at Vuebotics Corp., previously worked at Hughes Industrial Products Div., where he developed machine vision systems. Prior to Hughes, he was involved in integrated circuit design for Motorola.*

Reprinted from Proceedings of SPIE, Vol. 548, *Applications of Artificial Intelligence II* (1985)
Editor, J.F. Gilmore, © The Society of Photo-Optical Instrumentation Engineers

Robot Vision for Depalletizing Steel Cylindrical Billets

by Arthur V. Forman, Jr.

Abstract

A vision system is described for depalletizing steel cylindrical billets for a forging application. An algorithm for accurately locating and measuring the billets is described in some detail. Highlights of this discussion include an algorithm for adaptive threshold selection to accommodate changing image brightnesses and a special robot calibration procedure which enables inference of depth using only a single camera view combined with prior knowledge about the scene. Experimental results are presented which show that the system provides accurate measurements in spite of poor, inconsistent contrast.

Introduction

This paper describes the design and implementation of a vision system for depalletizing cylindrical metal billets in an industrial environment. Figure 1 depicts the problem. A pallet full of billets is placed under the vision station using a fork lift truck. A cylindrical coordinate robot must then remove the metal billets from the pallet and place each one in the rotary furnace. The vision system must accurately locate each billet so that the robot's gripper can grasp the billet and position it properly in the furnace. The objective for the vision system is to locate each billet within a tolerance of one eighth of an inch on the four-foot wide pallet (0.26 percent of the field of view). The vision system must also measure the length, width, and orientation of each billet. After the billet has been heated in the furnace, the robot removes the hot billet from the furnace and places it into a forging machine which pounds the metal into a turbine blade preform. The entire process takes about two minutes for a single billet. The vision system has approximately 20 seconds to complete its computations.

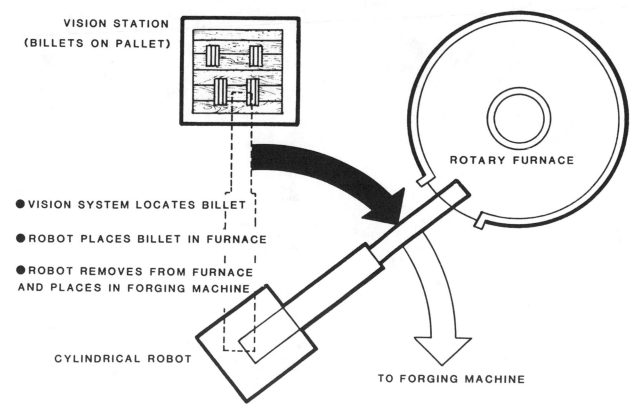

VISION STATION
(BILLETS ON PALLET)

ROTARY FURNACE

● VISION SYSTEM LOCATES BILLET

● ROBOT PLACES BILLET IN FURNACE

● ROBOT REMOVES FROM FURNACE
AND PLACES IN FORGING MACHINE

CYLINDRICAL ROBOT

TO FORGING MACHINE

Figure 1. Overview of depalletizing work cell.

Figure 2 is a photograph of the billets on a pallet which illustrates some of the issues which make this problem challenging. The pallet is typical of pallets found in a forging environment. In addition to the usual factory abuse that a pallet is likely to encounter, (e.g., nicks, smudges, dirt) the pallets in this forging area are constantly exposed to a fallout of fine black graphite dust which permeates the air and eventually covers everything in the cell. The pallets thus tend to become blacker and blacker with time. This might be used to advantage by the vision system were it not for the cylindrical shape of the metal billets to be measured.

The dominant characteristics of these objects are the bright specular stripes that run the length of each billet. These specular stripes provide high contrast and are sufficiently reliable to be used to measure the y-coordinate, orientation, and length of the cylinder. However, the edges on the long sides of the billet are not as well defined. As the line of sight from the camera to the billet surface approaches the tangent to the billet surface the brightness falls off until there is almost no contrast between the billet and the background. This makes repeatable extraction of the x-location and the width of the billet a difficult proposition. To further exacerbate the situation, the ambient light is bright and unpredictable. A large bay door near the work cell opens periodically, exposing the work cell to varying intensities of indirect sunlight.

Figure 2. Cylindrical billets on a typically dirty pallet.

Earlier works[4,5] have emphasized the overall solution to the problem of automating the activities in a forging cell. This paper focuses on the design of the vision system for the forging cell. In the paragraphs that follow, a vision system for depalletizing metal billets will be described with special emphasis on the image processing and algorithms to support this application. Experimental results will be presented that illustrate the accuracy that can be achieved when using this system for this application. Finally, conclusions and plans for future work will be presented.

System configuration

The solution to this problem, like the solution to any machine vision problem, consists of more than just processing hardware and software. In order for any gray-scale vision system to work effectively, it must be presented with an image that contains sufficient information to perform the required functions. In most cases, this means there must be adequate contrast between the part to be measured and its background. Contrast is achieved through properly selected lighting, camera, lensing, and fixturing[3]. The system configuration must provide an image of sufficient quality for the vision system and still satisfy all of the constraints imposed by the other elements of the work cell.

The simulation configuration consisted of flourescent tubes illuminating the pallet from the sides, a Sony XC-37 (491x384) CCD camera, and a 12.5mm C-mount lens mounted four feet above the pallet (See Figure 3). Sodium vapor lights are affixed to the outside of the frame of the vision system to simulate the ambient light in the forging area. The pallet is a common wooden pallet provided by the customer. (Ultimately, the configuration will include a hood and filters to suppress ambient light, will use standardized pallets, and will have the camera placed five feet above the pallet.)

This configuration of the system produces an image which enables measurement of the orientation, length, and y-location of each billet, because the bright specular stripes on each billet are brighter than the dirty wooden pallet. Because the dark shadows between the slats are darker than any points on the billets, it is possible to use these shadows to measure the width and x-location of each billet. This is the basis for the algorithm processing.

Figure 3. Vision station simulation and Opti-Vision™ system.

Algorithm processing

This section includes an overview of the algorithm flow, a description of the Opti-Vision™ image processing hardware and software, and a description of the algorithms for adaptive parameter adjustment and for accurately locating and measuring the billets.

Algorithm Flow Diagram

The algorithm flow diagram is presented in Figure 4. After the vision system receives a request from the robot, it will locate the top, left-most billet on the pallet by automatically computing a threshold to isolate the specular reflections on the billets. A window is defined around the first billet, and the height, y-location, and orientation of the billet are measured using the specular stripes. The automatically determined threshold is then reset to outline the dark cracks that fall within the window. The vertical edges on these cracks are used to measure the width and x-location of the billet. Using calibration data obtained earlier, the pixel measurements are converted to world coordinates for the robot, and these coordinates are transmitted to the robot. The robot then removes the billet, and the vision system waits for the next request.

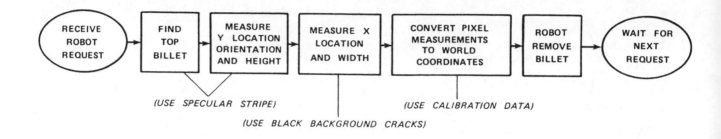

Figure 4. Algorithm processing flow for depalletizing metal billets

Image processing

Figure 5 depicts the hardware that is used for the algorithm and image processing. The Opti-Vision™ system is a gray-scale vision system which is centered around a fast pipeline image preprocessor and a general purpose Intel 8086/7 algorithm processor. The preprocessor performs image processing functions at near-video rates and outputs line segment data to the algorithm processor. The line segment data consists of the positions and orientations of line segments that coincide with with the edges, or boundaries, of parts in the scene. The preprocessor essentially outlines the parts by grouping edges into line segments. The preprocessor is programmed through specification of 39 software-selectable switches and parameters. The algorithm processor then operates on the line segment data to perform the desired functions. The algorithm processor is typically programmed in the PL-M™ programming language. For more detailed information about this system, see Reference[1].

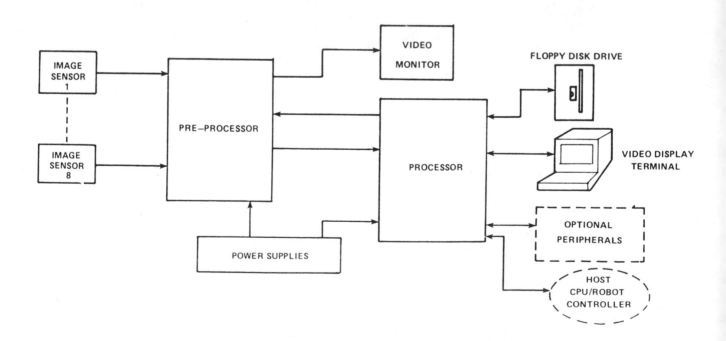

Figure 5. Opti-Vision™ image processing hardware: system overview.

Figure 6 illustrates the processing options from Opti-Vision™ that are used in this depalletizing application. The image input is standard RS-170 format interlaced video from the Sony XC-37. The image is first digitized into a 480-line array (512 samples per line, 6 bits per pixel). A histogram is then extracted from this image so that thresholds may be autonomously computed for extracting the bright specular and dark shadow regions in the scene. (The algorithm for automatic parameter adjustment is discussed later in this report.) A series of 3-pixel vertical and horizontal mean and median filters is applied to enhance the image. The effect of these filters is to remove high-frequency noise while simultaneously preserving edges - a compromise that can only be achieved using nonlinear filters like the median filter. These same filters remove unwanted texture from the image of the dirty pallet. The automatically computed thresholds are then applied to the enhanced image using the dual-level slicer. The resulting thresholded image is then passed through a Roberts cross gradient computation[1], which approximates the magnitude and direction of the gradient of the image at each pixel. (The system permits computation of the gradient of the gray-scale image as well, if desired.) The gradient directions are quantized into eight values. These directions are important because they are used to remove unwanted horizontal lines during the width computations. The gradient image is then thinned and thresholded to produce a thinned edge map. Points that belong to the boundary of a thresholded region (i.e., edge pixels) retain their gradient magnitude and direction, while all other points are zeroed. The edge pixels are then grouped together into line segments based on the connectivity of the edge points and on the compatibility of the gradient directions. The resulting line segments represent a tremendously compressed version of the image. The line segment data are the only data used in the "locate" and "measure" algorithm software.

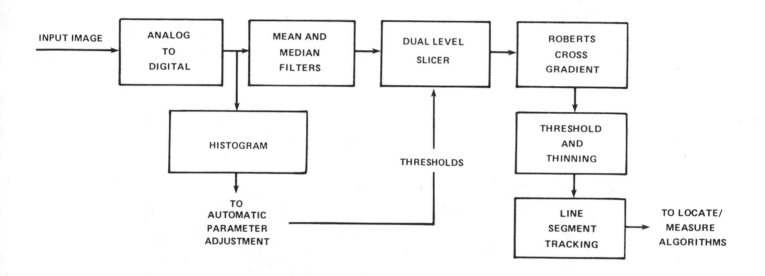

Figure 6. Image processing for depalletizing metal billets.

All of the aforementioned functions (except for the histogram) are performed in the special purpose pipeline hardware preprocessor and are accomplished at a rate of one microsecond per pixel, or .25 seconds to process a 480X512 image. An additional 0.7 seconds are required to extract the histogram. The rest of the processing is accomplished in the algorithm software.

Adaptive Parameter Adjustment

Because the average brightness of the image is constantly changing, either as a result of changing ambient light or as a result of the camera's AGC response to the removal of billets, a fixed brightness threshold will not suffice. An adaptive procedure for preprocessor threshold adjustment is required. The approach used in this application is based on the histogram of the image and assumes that the histogram is trimodal, as depicted in Figure 7. A gray value of zero represents bright white and a gray value of 63 represents dark black. The first (brightest) peak is caused by the bright specular region on each billet. As billets are removed, this peak will become smaller. (The absence of a bright mode is one indicator that there are no parts on the pallet.) The highest peak occurs at an intermediate gray value and is generated primarily by the wooden pallet. Some of this peak is also caused by the intermediate gray values on each billet. The third (darkest) peak is the result of the dark shadows between the wooden slats.

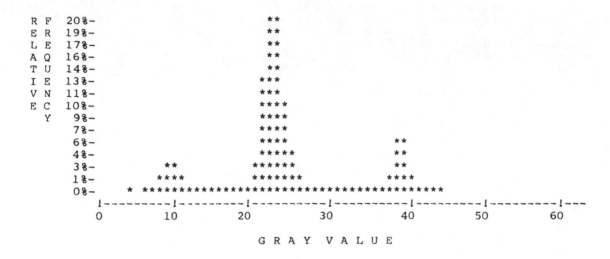

```
R F  20%-                        **
E R  19%-                        **
L E  17%-                        **
A Q  16%-                        **
T U  14%-                        **
I E  13%-                        ***
V N  11%-                        ***
E C  10%-                        ****
  Y   9%-                        ****
      7%-                        ****
      6%-                        ****              **
      4%-                        *****             **
      3%-            **          ******            **
      1%-          ****          *******          ****
      0%-   *  *********************************************
   !---------!---------!---------!---------!---------!---------!---
   0        10        20        30        40        50        60

              G R A Y   V A L U E
```

Figure 7.   Trimodal histogram extracted from image of metal billets lying on
             a wooden pallet.

In order to isolate the bright specular stripes, a threshold must be found that separates
the first bright peak from all other brightnesses in the scene. To isolate the dark shadow
regions, a threshold must be identified that isolates the third (darkest) peak from all
other pixels in the image.   The procedure is as follows. First the histogram of the image is
smoothed using a running average. Then the brightest local maximum, the darkest local maximum, and the highest local maximum are located.   These correspond to the specular regions,
the dark shadows, and the wooden pallet, respectively. Next the valleys are located between
these three peaks.   The valleys are the proper locations for the dark and bright thresholds.
The valley is identified either as an absolute minimum between peaks or by the change of
slope nearest the bright or dark peak.

The results from this algorithm are illustrated in Figures 8 and 9.   These figures show
the outlines of the regions that are generated after applying the automatically-determined
bright and dark thresholds, respectively, to the enhanced image.

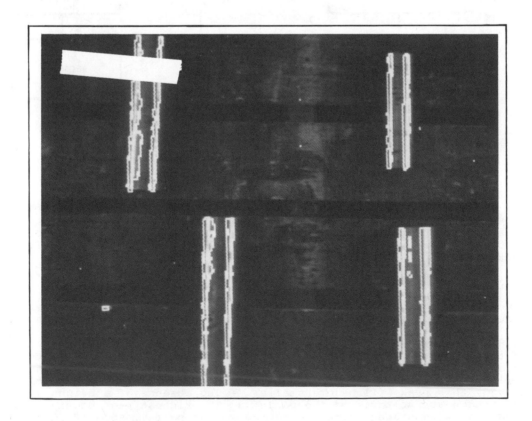

Figure 8.   Opti-Vision™ outlines of specular regions on each metal billet.
             (Threshold automatically selected.)

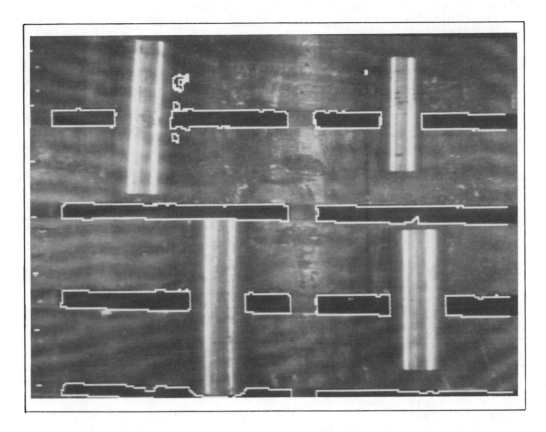

Figure 9. Opti-Vision™ outlines of dark cracks between the wooden slats on the pallet. (Threshold automatically selected.)

Measuring and locating parts

The outlines of the thresholded regions are the basis for the remaining calculations. First the outlines of the specular regions (See Figure 8.) are considered. A connectivity algorithm[6] is applied to a reduced-resolution version of the outlines to find the first billet. Using a reduced-resolution image for establishing connectivity has two benefits. First, small gaps in the outline are ignored so that all of the pixels on a part's boundary are properly connected. Second, the algorithm speed improves by a factor of 4 each time the resolution is halved. The results of the connectivity analysis at the reduced resolution are mapped back to the high resolution line segments so that line segments belonging to the outline of the first billet's stripes are flagged.

The first and second order moments[7] are computed from this outline, and the y-centroid and orientation are saved for later conversion to world coordinates. The length is obtained by measuring the distance from the minor axis of the specular outline to each point on the outline and by noting the points above and below the axis which are farthest away from the axis.

The width is computed using the major axis of the specular outline and the outlines of the dark background cracks. (See Figure 9). First the horizontal lines and any lines near the top and bottom of the billet are removed. Then the distance from the major axis of the specular outline to each vertical line segment is computed. The width is determined by the points nearest to the left and to the right of the major axis. The x-centroid is then computed as the midpoint of these two projected distances.

The expectation for these measurements is that the x-centroid and y-centroid will be very accurate, while the length, width, and orientation measurements will be less accurate. This is due to the fact that a single bad pixel can ruin the length and width measurements, while the centroid computations in this algorithm are only one fourth as sensitive to a single bad pixel.

Robot calibration

The final step in the algorithm processing is to convert pixel measurements into world coordinates. The problem is to find a function "f" that accurately predicts world coordinates from image measurements.

What is the form of the function "f"? Image formation is commonly modeled as a perspective transformation of the three-dimensional world onto a two-dimensional image plane[2]. This transformation discards the depth information about the scene. The perspective transformation is an ideal that is rarely realized in practice - lens distortions exist (especially near the edge of the field of view) which alter this relationship between world coordinates and image coordinates[8]. Distance to points in the scene cannot be computed from a single image unless some prior knowledge is provided about the scene. In this application the depth information might be recovered by observing that cylinders lie on a planar surface and have a circular cross section and therefore the z coordinate for the robot may be related to the width of the image of the billet as computed by the algorithm. It is assumed that the function f is a smooth function, and therefore that a Taylor series expansion of f is a good approximation to the true function over the field of view, particularly if the calibration points are well-chosen.

After trying a number of well-known functions f, the following function was selected because it provides a mechanism for measuring the world z coordinate along with the world x and y coordinates:

$$x_r = a_0 + a_1 x_i + a_2 y_i + a_3 w_i,$$
$$y_r = b_0 + b_1 x_i + b_2 y_i + b_3 w_i, \text{ and} \tag{1}$$
$$z_r = c_0 + c_1 x_i + c_2 y_i + c_3 w_i,$$

where

x_i = image x centroid,
y_i = image y centroid,
w_i = width of image of billet,

x_r = world x coordinate,
y_r = world y coordinate, and
z_r = world z coordinate.

This transformation accurately predicts the world coordinates (x_r, y_r, z_r) from the three image measurements (x_i, y_i, w_i). Errors due to lens distortions are not predicted by this function, but they have been observed to be small, as discussed later in this report.

The robot calibration, then, reduces to a problem of finding the twelve coefficients a_0-a_3, b_0-b_3, and c_0-c_3. The procedure that is required is to have the robot place a billet in the field of view and note the world coordinates of the part. The robot arm then retracts and the image measurements are computed using the algorithms described earlier. For every part that the robot places in front of the vision system during calibration, a set of three equations are generated. After the robot has placed the billet in four positions, a system of twelve equations in twelve unknowns has been generated. This system can be solved for the twelve calibration coefficients. However, better results can be obtained by using more calibration points. A least-squares solution[2] can be applied to the resulting over-constrained system of equations. It should be emphasized that the computation of the calibration coefficients is performed offline prior to the actual depalletizing operation, and so the time required to perform these computations is not a major concern. Once the coefficients have been computed, it is a quick and simple process to convert the image measurements to world coordinates by using Equations (1).

Experimental results

An experiment was conducted in an effort to measure the accuracy of the vision system. In lieu of a robot, the "true" measurements were obtained using a mechanical measurement system consisting of fine nylon monofilament line, sliding metal clips, measuring tapes, and springs affixed to the lower part of the frame of the vision station. This system is estimated to provide measurements which are accurate to within one sixteenth of an inch. Obviously, this imposes a limit on the accuracy with which the robot can be calibrated and on the accuracy with which the measurments of the vision system can be verified.

The robot was calibrated using four billets placed at five locations on the pallet. The billets varied in length from 8.16 to 10.88 inches and in diameter from 3.00 to 4.03 inches. Each location was used twice in an effort to reduce the effect of measurement errors on the calibration procedure, resulting in a total of ten calibration measurements. After the robot was calibrated, the billets were left in their calibration positions and their mechanically-measured location (i.e., the "true" location) was compared to the vision system's computed billet location. These results are tabulated in Table 1. They are typical of results that were obtained over several days of experimentation. The residual errors in Table 1 are so small as to be attributable to the errors introduced by the mechanical measurement system. When the billets were moved to new locations on the pallet, the locations provided by the vision system were also very accurate, again, allowing for the errors introduced by the mechanical measurement system. The errors of the vision system were too small to be measured reliably. The vision system must be integrated with a robot if the true accuracy of the system is to be known.

	CASE		X (INCHES)	Y (INCHES)	Z* (INCHES)	ORIENTATION (DEGREES)	LENGTH (INCHES)

TABLE I

True (mechanically measured) vs. computed (vision system) measurements at the calibration points.

CASE		X (INCHES)	Y (INCHES)	Z* (INCHES)	ORIENTATION (DEGREES)	LENGTH (INCHES)
1	TRUE	12.19	17.31	2.0	-----	10.81 - 10.88**
1	COMPUTED	12.20	17.22	1.99	92.14	10.78
1	COMPUTED	12.20	17.22	1.98	92.08	10.84
2	TRUE	29.31	29.13	2.0	-----	12.31
2	COMPUTED	29.34	29.05	1.93	90.63	12.24
2	COMPUTED	29.30	29.08	1.97	91.03	12.24
3	TRUE	8.31	30.75	1.5	-----	8.16
3	COMPUTED	8.34	30.71	1.50	88.90	7.98
3	COMPUTED	8.32	30.70	1.47	88.68	7.98

  * Width of billet is two times the Z coordinate.
  ** Billet was not cut perpendicular to cylinder axis.

As expected, greater accuracy has been observed in the centroid measurements (x and y position) than in the length measurements. The limiting factors in the accuracy of the billet locations are the resolution of the camera and the accuracy of the calibration. The limiting factor in the measurements of length and width appears to be contrast - slight (1-pixel) inconsistencies in the images of the objects are creating the greatest error in these measurements.

Conclusions and plans

The accuracy for locating centroids appears to be approximately one pixel, and perhaps better. The true accuracies of the computed world coordinates for this approach cannot be determined precisely until the system is integrated with a robot.

Accuracy could be improved in a number of ways. For example, instead of using the pixels that are nearest or farthest from an axis to measure length and width (a very noise-sensitive process), a curve fitting procedure could be applied to locate the sides of the part, potentially providing sub-pixel accuracy in the measurements. Accuracy might also be improved by taking a sequence of images and measurements and averaging the results. Greater accuracy could also be obtained by working with a smaller field of view, either by using multiple cameras, by moving cameras, or by having the robot place each billet in a separate inspection area. Calibration errors might be reduced by adding the second-order terms to the Taylor series expansion for the Equations (1) for X_r.

The next step in this experiment will be to integrate the vision system with a robot and to perform tests to measure the statistical fluctuations in the measurements. (The customer has only recently installed the robots in the automated forging cell). Algorithm robustness (insensitivity to broad variations in the appearance of billets and pallets) and algorithm repeatability are important characteristics that can only be confirmed with more experimentation. However, the preliminary results indicate that this approach has potential for providing accurate billet locations for robotic depalletizing.

Bibliography

1. Automation Intelligence, Inc. Opti-Vision™ M-100 User's Guide. AII, 1200 West Colonial Drive, Orlando, Fla., 32804-7194, December 1984.

2. Ballard, Dana H. and Christopher M. Brown. Computer Vision. Prentice-Hall, Inc., Englewood Cliffs, N.J., 1982. pp. 477-489.

3. Hopwood, Ronald K. "Design Considerations for a Solid-State Image Sensing System". Society of Photo-Optical Instrumentation Engineers (SPIE), Vol. 230 (Minicomputers and Microprocessors in Optical Systems), 1980. pp. 72-82.

4. Miller, Paul C. "Software Basics for a Robotic-Cell Story". Tooling and Production, September 1984. pp 75-78.

5. Perkins, James M. "Three Robot Extrusion Workstation". Robotics International of the Society of Manufacturing Engineers, Dearborn, Michigan. Technical Paper MS82-132, 1982.

6. Perkins, W. A. "Area Segmentation of Images Using Edge Points". IEEE Transactions on Pattern Analysis and Machine Intelligence, Volume PAMI-2, No. 1., January 1980. pp. 8-15.

7. Teague, Michael Reed. "Image Analysis Via the General Theory of Moments". Journal of the Optical Society of America, Volume 70, Number 8., August 1980. pp. 920-930.

8. Wolfe, William L. and George J. Zissis. The Infrared Handbook. Prepared by IRIA Center, Environmental Research Institute of Michigan, for the Office of Naval Research, Department of the Navy, Washington, D.C. 1978. pp. 8-1 to 8-39.

CHAPTER 5

FLEXIBLE MANUFACTURING SYSTEMS

Presented at the CASA/SME AUTOFACT 5 Conference, November 1985

Justification of Flexible Manufacturing Systems Using Expert System Technology

William G. Sullivan
University of Tennessee

and
Steven R. LeClair
U.S. Air Force

Introduction

There is growing concern that traditional financial evaluation criteria, such as return on investment (ROI) and net present worth (NPW), are not capable of dealing adequately with proposed investments in advanced manufacturing technologies [2,6,12,17,19]. These technologies typically involve the integration of a firm's resources and data bases, and they have far-reaching implications for the firm's competitiveness and even survival. Some of these implications and concerns can be reduced to monetary criteria such as the familiar ROI, while many others evade quantification in terms of dollars and cents.

It is the objective of this paper to present an expert system, called XVENTURE, which has been developed to evaluate the justification of a generic flexible manufacturing system (FMS). This expert system makes use of six groupings of issues that are vitally important to investment decisions in strategic investments involving advanced manufacturing technology and links them together into a rule base representing the expertise of a single expert. It should be noted that XVENTURE makes use of a traditional financial evaluation criterion (net present worth) in addition to five other issues that tend to be "irreducibles" in the process of balancing benefits and costs for a proposed capital investment. The six categories of issues are shown in Figure 1.

Figure 1. Six Categories of FMS Justification Issues

1. Managing today's investment options for future growth [9]
2. Matching the firm's strategic business plan with its technology plan [18]
3. Modifying accounting practices to reflect changes in cost patterns [8]
4. Dealing with uncertainties in the business environment and in the technology itself [1]
5. Considering benefits of improved quality, flexibility, throughput times and inventory positions [5]
6. Quantifying a proposed project's profitability/liquidity with traditional financial evaluation techniques [3]

The reasons for developing XVENTURE are basically twofold: (1) to offer an alternative to the sole use of traditional financial criteria which are believed to be inadequate for representing the long-range implications of integrated manufacturing systems, and (2) to demonstrate that an approach to combining the tactical and strategic dimensions of FMS justification studies can be programmed into an expert system and utilized as the basis for a multiple expert system from which learning will occur.

161

The Controversy Over ROI and NPW

Some authors advocate rejection of traditional financial evaluation techniques in FMS justifications [6] while other writers state that these techniques are suitable in such studies but that they have been incorrectly applied [11]. There is some validity to the notion that we should learn to use better what we already have before "developing a better mousetrap." By obtaining better cost and revenue estimates and accurately defining the cash flow implications of the "do nothing" alternative relative to an FMS, a more valid assessment of incremental benefits and costs can be incorporated into the ROI or NPW calculation. However, the undeniable truth is that many benefits of an FMS, such as improved throughput time and flexibility, cannot be reduced to dollars and still be credible to corporate decision makers. Instead, decision makers typically are provided with studies that show "the present worth of easily quantified savings of an FMS is less than present worth of costs, but it is believed the present worth of intangible benefits far exceeds that of intangible costs. Therefore, the investment should be made."

An Expert System Approach

The justification of computer integrated manufacturing systems such as an FMS requires numerous disparate aspects of the investment opportunity to be simultaneously considered. Thus, long- and short-term perspectives must be placed beside profitability, manufacturing flexibility, market uncertainty, and "do-ability" considerations so that a wide array of important monetary and nonmonetary factors are evaluated in arriving at a decision. It is our contention that an expert system is a promising tool for integrating these critical concerns into a qualitative and believable approach for justifying a flexible manufacturing system.

An expert system (ES) is "a knowledge-intensive program that solves problems normally requiring human expertise" [7]. XVENTURE is a turn-key ES developed and run on an IBM (or compatible) personal computer with 128K of memory using an MS-DOS operating system. An ES development system called EXPERT4 was utilized to create XVENTURE. EXPERT4 is programmed in FORTH and is a recent version of EXPERT2 distributed by Mountain View Press of Mountain View, California [14].

The advantage of using an ES such as XVENTURE for justification of an FMS is that it affords a decision maker available expertise with above average ability to analyze strategic, nonquantifiable decision attributes that lend themselves well to heuristic analysis. Furthermore, an ES preserves otherwise perishable human expertise and reduces the risks of poor human performance [4,20].

To address the widely acknowledged need for holistic approaches to FMS justification, it is clear that the total impact of factory automation on a particular firm must be considered. A comprehensive view of advanced manufacturing systems must be taken to include interactions among all components considered in the strategic decisions of a company. Developing XVENTURE around six basic issues represents a unique but viable approach to enabling a more comprehensive view of FMS justification.

The ES illustrated in this section was designed through two-way communication between one author who acted as an expert in FMS justification and the second author who served as a knowledge engineer. The expert revealed fragments of his thought process as he considered the six issues in arriving at "GO, DEFER, and NO GO" decisions concerning justification of a typical FMS. Concurrently, the knowledge engineer listened carefully for problem-solving elements in the dialog and attempted to create a description of the way the expert arrived at a "GO, DEFER or NO GO" determination for a particular scenario of issue outcomes in the FMS justification process. Then by using the ES development program (EXPERT4), a paradigm comprised of IF-THEN rules was constructed that strived to mimic the expert for different combinations of issue outcomes. Further testing and improvement were needed to "fine tune" the rule base of XVENTURE so that it produced logically defensible results.

The six statements shown in Figure 2 are based on issues presented in Figure 1 and they appear in the order shown to the user of XVENTURE. The computer allows the user to select the most appropriate response by moving the cursor to the desired location with the "space" bar and then tapping the "return" key.

To facilitate understanding XVENTURE and how it is utilized, the following example session is provided. The session focuses on four rules (a, b, c and d) in the XVENTURE rule base identified in Table 1 and illustrated in Figure 3.

It can be seen from Table 1 that each rule requires particular responses to the questions represented in Figure 2. The cursor ($>$) symbols in Figure 3 indicate the appropriate response for all questions regarding the four rules selected from Table 1. Note from Table 1 that a response of "technology window will be open for years and is not proprietary" to the first question (Figure 2) reduces the possible rules for further consideration to rule b in Figure 3 since rules a, c and d require "the technology is immediately available for price" response.

The above example session was limited to just four rules. When considering the entire rule base, the XVENTURE rule interpreter continues the process of elimination and selects those rules that are still applicable after each response. Even though rule b from Figure 3 may be the only rule (other than the collapse rule) still applicable, XVENTURE requires the appropriate remaining responses from questions in Figure 2 to deduce the hypothesis that rule b produces; that is, VENTURE IS JUSTIFIABLE - GO. If the appropriate responses were not provided by the decision maker, then XVENTURE would "collapse" the rule base and conclude REJECT/ABANDON VENTURE - NO GO. The collapse function acts as a default, i.e. if all other rules fail, the rule interpreter finds the collapse function to end the session.

Figure 2. Basic Issues and Responses in XVENTURE

1. Today's investment options must be effectively managed for future growth. Which of the following best describes the investment opportunity being evaluated?

    1 1* The technology is immediately available for a price

    2 1   Technology window will be open for years and is not proprietary

    3 1   Technology window will be open for years and is proprietary

2. The firm's technology plan should match its business plan for the next five years, otherwise success of new technology is questionable. Select one of the following which best reflects your technology/business match regarding this investment.

    1 2   The two plans are not likely to ever agree

    2 2   They do not agree now but may in 1 - 2 years

    3 2   They do agree now but may not in 1 - 2 years

    4 2   They watch perfectly with no changes foreseen

3. Consider your firm's present cost accounting system. How would you rate prospects for a future accounting system that considers direct and indirect labor costs to be fixed and a flexible manufacturing system investment to be a variable cost based on utilization?

    1 3   Absolutely not possible during the foreseeable future

    2 3   Management is reluctant but will give the concept a fair review

    3 3   Support is excellent and such a system is being developed.

4. A complex investment tends to be strategic and long-term in nature, while a simple investment is tactical and short-term in outlook. Choose which statement best describes the investment's complexity and technological/environmental uncertainty.

    1 4   Investment is complex and there is considerable uncertainty

    2 4   Investment is complex and there is moderate to no uncertainty

    3 4   Investment is simple and there is moderate to no uncertainty

5. The investment's promotion of more flexible capacity, better product quality and improved productivity is essential. Select the most appropriate statement regarding the investment opportunity's impact on the above.

    1 5   There is very little improvement in the long-term

    2 5   The investment makes marginal improvements

    3 5   The investment makes significant long-term improvements

6. Discounted cash flow techniques indicate an investment's long-term profitability and are used to evaluate the monetary attributes of an investment. Calculate the after-tax present worth of the proposed FMS and choose one of the following.

    1 6   The present worth is greater than or equal zero

    2 6   The present worth is less than zero

* Responses are coded as they appear in Table 1.

TABLE 1. XVENTURE RULE BASE

```
Figure 9(a) --> 1 1 IF 4 2 AND 2 3 AND 2 4 AND 3 5 AND 2 6 AND 7 THENHYP
                1 1 IF 3 2 AND 2 3 AND 3 4 AND 2 5 AND 2 6 AND 7 THEN
                2 1 IF 2 2 AND 2 3 AND 2 4 AND 3 5 AND 2 6 AND 7 THEN
                2 1 IF 3 2 AND 2 3 AND 2 4 AND 3 5 AND 2 6 AND 7 THEN
Figure 9(b) --> 2 1 IF 4 2 AND 2 3 AND 2 4 AND 3 5 AND 2 6 AND 7 THEN
                3 1 IF 1 2 AND 3 3 AND 3 4 AND 3 5 AND 2 6 AND 7 THEN
                3 1 IF 2 2 AND 2 3 AND 3 4 AND 3 5 AND 2 6 AND 7 THEN
                3 1 IF 3 2 AND 2 3 AND 2 4 AND 3 5 AND 2 6 AND 7 THEN
                3 1 IF 4 2 AND 2 3 AND 2 4 AND 3 5 AND 2 6 AND 7 THEN
                1 1 IF 4 2 AND 3 3 AND 1 4 AND 3 5 AND 2 6 AND 7 THEN
                1 1 IF 3 2 AND 3 3 AND 1 4 AND 3 5 AND 2 6 AND 7 THEN
                2 1 IF 2 2 AND 3 3 AND 2 4 AND 3 5 AND 2 6 AND 7 THEN
                2 1 IF 4 2 AND 3 3 AND 1 4 AND 3 5 AND 2 6 AND 7 THEN
                3 1 IF 2 2 AND 3 3 AND 2 4 AND 3 5 AND 2 6 AND 7 THEN
                3 1 IF 3 2 AND 3 3 AND 1 4 AND 3 5 AND 2 6 AND 7 THEN
                3 1 IF 4 2 AND 3 3 AND 1 4 AND 3 5 AND 2 6 AND 7 THEN
                3 1 IF 4 2 AND 3 3 AND 2 4 AND 1 5 AND 2 6 AND 7 THEN
                3 1 IF 4 2 AND 3 3 AND 3 4 AND 1 5 AND 2 6 AND 7 THEN
                1 1 IF 3 2 AND 2 3 AND 3 4 AND 3 5 AND 2 6 AND 7 THEN
                3 1 IF 3 2 AND 2 3 AND 3 4 AND 3 5 AND 2 6 AND 7 THEN
                3 1 IF 4 2 AND 2 3 AND 3 4 AND 3 5 AND 2 6 AND 7 THEN
                1 1 IF 4 2 AND 3 3 AND 2 4 AND 3 5 AND 2 6 AND 7 THEN
                1 1 IF 4 2 AND 3 3 AND 3 4 AND 2 5 AND 2 6 AND 7 THEN
                1 1 IF 4 2 AND 3 3 AND 3 4 AND 3 5 AND 2 6 AND 7 THEN
                1 1 IF 3 2 AND 3 3 AND 2 4 AND 3 5 AND 2 6 AND 7 THEN
                1 1 IF 3 2 AND 3 3 AND 3 4 AND 2 5 AND 2 6 AND 7 THEN
                1 1 IF 3 2 AND 3 3 AND 3 4 AND 3 5 AND 2 6 AND 7 THEN
                2 1 IF 2 2 AND 3 3 AND 3 4 AND 3 5 AND 2 6 AND 7 THEN
                2 1 IF 4 2 AND 3 3 AND 2 4 AND 3 5 AND 2 6 AND 7 THEN
                2 1 IF 4 2 AND 3 3 AND 3 4 AND 2 5 AND 2 6 AND 7 THEN
                2 1 IF 4 2 AND 3 3 AND 3 4 AND 3 5 AND 2 6 AND 7 THEN
                3 1 IF 3 2 AND 3 3 AND 2 4 AND 3 5 AND 2 6 AND 7 THEN
                1 1 IF 4 2 AND 2 3 AND 3 4 AND 3 5 AND 2 6 AND 7 THEN

2 1 IF 2 2 AND 2 3 AND 3 4 AND 3 5 AND 2 6 AND 7 THEN
2 1 IF 3 2 AND 2 3 AND 3 4 AND 3 5 AND 2 6 AND 7 THEN
2 1 IF 4 2 AND 2 3 AND 3 4 AND 3 5 AND 2 6 AND 7 THEN
3 1 IF 3 2 AND 2 3 AND 3 4 AND 2 5 AND 2 6 AND 7 THEN
3 1 IF 4 2 AND 2 3 AND 3 4 AND 2 5 AND 2 6 AND 7 THEN
1 1 IF 4 2 AND 3 3 AND 2 4 AND 2 5 AND 2 6 AND 7 THEN
1 1 IF 4 2 AND 3 3 AND 3 4 AND 1 5 AND 2 6 AND 7 THEN
1 1 IF 3 2 AND 3 3 AND 2 4 AND 2 5 AND 2 6 AND 7 THEN
1 1 IF 3 2 AND 3 3 AND 3 4 AND 1 5 AND 2 6 AND 7 THEN
2 1 IF 2 2 AND 3 3 AND 3 4 AND 2 5 AND 2 6 AND 7 THEN
2 1 IF 4 2 AND 3 3 AND 2 4 AND 2 5 AND 2 6 AND 7 THEN
2 1 IF 4 2 AND 3 3 AND 3 4 AND 1 5 AND 2 6 AND 7 THEN
3 1 IF 2 2 AND 3 3 AND 3 4 AND 3 5 AND 2 6 AND 7 THEN
3 1 IF 3 2 AND 3 3 AND 2 4 AND 2 5 AND 2 6 AND 7 THEN
3 1 IF 3 2 AND 3 3 AND 3 4 AND 1 5 AND 2 6 AND 7 THEN
3 1 IF 3 2 AND 3 3 AND 3 4 AND 2 5 AND 2 6 AND 7 THEN
3 1 IF 3 2 AND 3 3 AND 3 4 AND 3 5 AND 2 6 AND 7 THEN
3 1 IF 4 2 AND 3 3 AND 2 4 AND 2 5 AND 2 6 AND 7 THEN
3 1 IF 4 2 AND 3 3 AND 3 4 AND 3 5 AND 2 6 AND 7 THEN
3 1 IF 4 2 AND 3 3 AND 3 4 AND 2 5 AND 2 6 AND 7 THEN
3 1 IF 4 2 AND 3 3 AND 3 4 AND 3 5 AND 2 6 AND 7 THEN
1 1 IF 4 2 AND 3 3 AND 3 4 AND 1 5 AND 1 6 AND 8 THENHYP Figure 9(c)
1 1 IF 3 2 AND 3 3 AND 3 4 AND 1 5 AND 1 6 AND 8 THEN
3 1 IF 4 2 AND 3 3 AND 3 4 AND 1 5 AND 1 6 AND 8 THEN
1 1 IF 4 2 AND 3 3 AND 3 4 AND 2 5 AND 1 6 AND 8 THEN
1 1 IF 4 2 AND 3 3 AND 3 4 AND 3 5 AND 1 6 AND 8 THEN
1 1 IF 3 2 AND 3 3 AND 3 4 AND 3 5 AND 1 6 AND 8 THEN
2 1 IF 4 2 AND 3 3 AND 3 4 AND 2 5 AND 1 6 AND 8 THEN
2 1 IF 4 2 AND 3 3 AND 3 4 AND 3 5 AND 1 6 AND 8 THEN
3 1 IF 3 2 AND 3 3 AND 3 4 AND 3 5 AND 1 6 AND 8 THEN
3 1 IF 4 2 AND 3 3 AND 3 4 AND 3 5 AND 1 6 AND 8 THEN
3 1 IF 4 2 AND 3 3 AND 3 4 AND 2 5 AND 1 6 AND 8 THEN
3 1 IF 4 2 AND 3 3 AND 3 4 AND 3 5 AND 1 6 AND 8 THEN
9 COLLAPSE <-- Figure 9(d)
```

LEGEND

7 ⇒ VENTURE IS JUSTIFIABLE - GO

8 ⇒ DECISION IS NO GO FOR NOW BUT COULD BE GO AT LATER DATE - DEFER

9 ⇒ REJECT/ABANDON VENTURE - NO GO

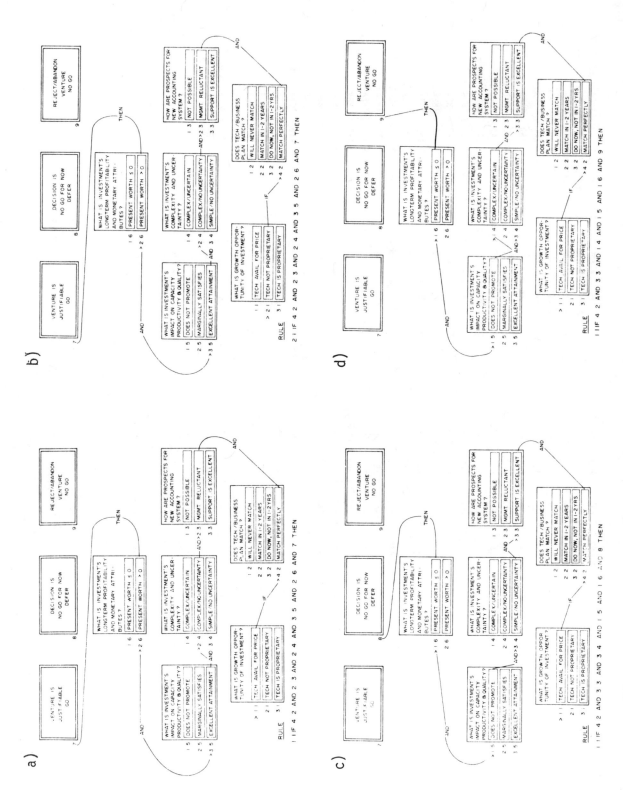

Figure 3. Illustrative XVENTURE Session Utilizing Four Rules from Table 1.

It should be noted that given the six XVENTURE questions and their multiple responses, a total of 648 rules is possible, yet the entire rule base shown in Table 1 consists of just 66 rules. The remaining 582 rules (of which rule d from Figure 3 is one) lead to the "collapse" (NO GO) conclusion, thus avoiding the writing of these 582 individual rules.

One final item of interest is that once a conclusion is reached, the user only needs to tap the space bar to reinitialize XVENTURE for a new session.

Comparison of XVENTURE with Problem Solving Algorithms

XVENTURE is typical of artificial intelligence (AI) approaches to problem solving—it relies on pattern matching and heuristic search to find a solution. The heuristics in the case of XVENTURE are rules (abstractions) organized to form a line of reasoning which represents the knowledge of an expert. An expert is defined as someone who is recognized as having special knowledge in solving specific tasks. But algorithmic approaches also use abstractions in the form of mathematical models. So, what is the difference between AI and algorithmic approaches?

Actually the differences are subtle but significant. First, AI approaches attempt to solve problems by representing knowledge based on previous experiences regarding the problem domain, whereas algorithms arrive at a solution by characterizing the problem so that it fits some solution technique structure. Secondly, because AI systems use heuristics, they behave much like humans in that they use recognizable patterns to reach a decision, whereas algorithms quantify the problem and compute a solution. When compared to algorithms, AI systems are more flexible, less sophisticated and better able to react to particular problem situations. But, it should be noted that AI systems must 'second-guess' all possible situations within a problem domain to be as encompassing as an algorithm in determining the best solution.

Although algorithms are expected to be more quantitative in nature and operate on more specific (richer) knowledge about the problem, they do not necessarily provide a "more correct" answer. Consider the problem of justifying a proposed FMS investment. Numerically coded approaches are practical for this type of problem, such as Morris' ranking and rating procedure [13] and Saaty's analytical hierarchy process [16]. However, when considering tasks such as capital investment analysis, there are various aspects of the problem which suggest that an AI approach is superior.

One aspect concerns the values assigned to the range of responses to each question. Responses associated with FMS investment questions are often dependent on other responses. By treating each question independently of the others, the synergistic effect of a particular set of responses could easily be overlooked.

Another area of concern is encountered when an FMS decision is based on the sum of numerical values above/below some minimally acceptable threshold value. Can a single threshold value realistically be the discriminator for all possible combinations of responses? A threshold value is often only an average or median score for the range of all possible values produced by an algorithm. Even if a more in-depth criterion were used, such numerical approaches begin to appear arbitrary and inadequate for providing sound advice in the vicinity of their threshold values.

Another point of comparison of AI with algorithmic approaches concerns ease of use and maintenance. Because problems are typically not static, approaches to solving them must be capable of adapting to changes in the environment. Also, changes in knowledge regarding how to solve a problem should be expected over time. In systems such as XVENTURE, the knowledge is in the form of rules which at any time can be changed by adding new rules, modifying existing rules or deleting rules that reflect either changes in knowledge or in the environment.

Current Research with Multiple Experts

In its present form, XVENTURE represents the knowledge of a single expert in the problem domain of FMS justification. Therefore, XVENTURE contains a single line of reasoning. Even though a single-expert, single-line-of-reasoning ES is the simplest to develop, it is rare. Most ESs typically represent the expertise of many experts who collectively contribute to a "pooled" knowledge base [15].

The conventional way to develop a multiple expert knowledge system is to seek consensus among the experts during knowledge acquisition (i.e. before building the knowledge base). The ES then results in a single line of reasoning which represents a blending of the expertise of multiple experts. But by blending expertise, the knowledge sometimes becomes illogical and error-prone because it is difficult to achieve consensus without awkward compromise of each expert's reasoning.

It is recognized that if ESs are to become accepted in the business world, the will have to accommodate multiple experts with potentially conflicting expertise. Most decisions encourage multiple conflicting inputs to establish appropriate priorities and evaluate alternatives for any given problem situation. For example, in evaluating an FMS one expert may consider issues focused on near-term quality improvements, another on long-term survival of the firm, and still another on cash-flow stability over the next 5 years. Priorities associated with these views may change because various investments affect the priorities differently.

Research currently being conducted by the authors addresses the above issues and actually takes advantage of multiple experts to enable learning and knowledge-base maintenance. The approach developed by LeClair [10] is discussed in the following paragraphs and is summarized in Figure 4.

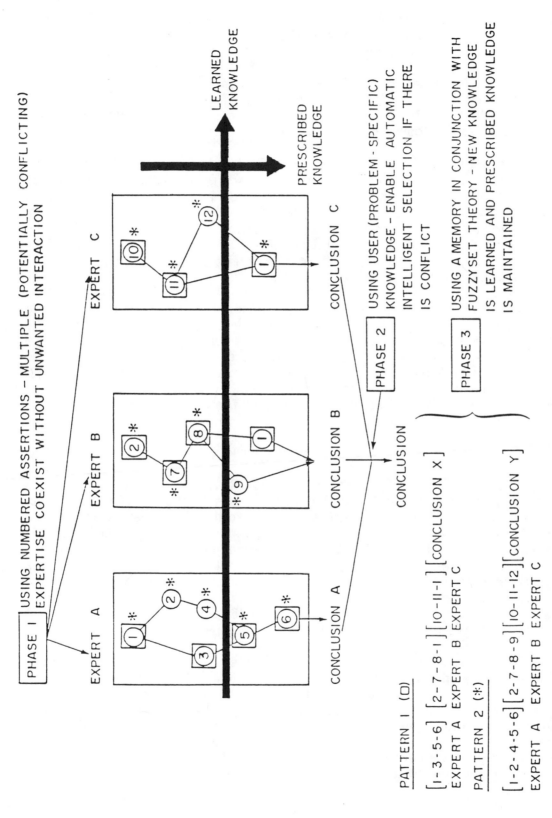

Figure 4. Design of a Multiexpert Knowledge System.

The first step in building a multiexpert (multiple-expert, multiple-line-of-reasoning) knowledge system is to enable the integration of multiple experts with potentially conflicting expertise in the same knowledge base. The problem is to accommodate each expert as a complete self-contained single-expert, single-line-of-reasoning system and at the same time enable the sharing of knowledge between experts. The sharing is such that the user is not confused by the questions and is asked any particular question only once regardless of how many experts may use the response.

The integration of multiple experts is handled such that the user can seek any single line of reasoning or be apprised of the conclusions from all experts. Thus the ES can operate in either a single expert mode or as a multiexpert system based on the desire of the user.

The second step is to enable the ES to recognize conflict between experts and provide the decision maker a means to evaluate the conflict such that a preferred line of reasoning can be selected. An architecture advocated by LeClair attempts to improve upon existing conflict resolution techniques by considering knowledge presented by the user regarding the particular problem (e.g., justification of an FMS) of interest. This new knowledge is used by the system to recommend the most appropriate line of reasoning and the corresponding conclusion.

The third step involves the use of multiple lines of reasoning to enable learning and knowledge-base maintenance. The ES uses the responses of the user coupled with either the consensus or selected (in the case of conflict) conclusion to form new patterns (or rules) which become learned knowledge. Such learned knowledge can then be used to modify the behavior of the ES. Multiexpert knowledge systems appear to be a fruitful area for further research in the justification of advanced manufacturing systems.

Concluding Remarks

The use of an ES, such as XVENTURE, is a novel approach to combining strategic and tactical considerations in the justification of an FMS. XVENTURE uses knowledge in the form of heuristics to resolve quickly and consistently a multidimensional FMS justification problem. It is recognized that the XVENTURE knowledge base is generic and not necessarily representative of any particular industry or company. But the significance of XVENTURE is that it demonstrates the feasibility and benefits of such an approach. Given the ease with which ESs are constructed and maintained, the XVENTURE rule base can be adapted and/or enhanced for any specific industry or company.

The authors firmly believe that multiexpert knowledge systems are the next logical step for expert systems in decision making environments. To ensure the continued improvement in quality and accuracy of advice is for that advice, ESs must be challenged. Thus, to foster such an environment, a multiexpert knowledge system with potentially conflicting expertise is needed. The end result will be an ES which not only learns and improves its performance but evolves over time and modifies its behavior to reflect changes in the knowledge base as well as changes in the environment.

References

1. Ayres, R. U. and Miller, S., "Robotics, CAM and Industrial Productivity," National Productivity Review, (Winter 1981-1982), pp. 42-60.

2. Curtin, F. T., "Planning and Justifying Factory Automation Systems," Production Engineering, (May 1984), pp. 46-51.

3. DeGarmo, P. E., W. G. Sullivan and J. R. Canada, Engineering Economy (Seventh Edition), MacMillan Publishing Co., New York (1984).

4. Gevarter, W. B., "Expert Systems: Limited But Powerful," IEEE Spectrum, (August 1983a), Vol. 20, pp. 39-45.

5. Gold, B., "CAM Sets New Rules for Production," Harvard Business Review, (November-December 1982), pp. 88-94.

6. Hayes, R. H., and David A. Garvin, "Managing As If Tomorrow Mattered," Harvard Business Review, (May-June 1982), pp. 71-79.

7. Hayes-Roth, F., "Knowledge-Based Expert Systems," Computer, (October 1984), pp. 263-273.

8. Kaplan, Robert S., "Yesterday's Accounting Undermines Production," Harvard Business Review, (July-August 1984), pp. 95-101.

9. Kester, W. Carl, "Today's Options for Tomorrow's Growth," Harvard Business Review, (March-April 1984), pp. 153-160.

10. LeClair, S., "A Multiexpert Knowledge System Architecture for Manufacturing Decision Analysis," Ph.D. Dissertation, Arizona State University, (May 1985).

11. Logue, Dennis E. and Richard R. West, "Discounted Cash Flow Analysis: A Response to the Critics," National Productivity Review, (Summer 1983), pp. 233-241.

12. Meyer, Ronald J., "A Cookbook Approach to Robotics and Automation Justification," Society of Manufacturing Engineers, Dearborn, Michigan, Report No. M582-192, (1982).

13. Morris, W. T., Engineering Economic Analysis, Reston Publishing Company, Reston, Virginia, (1976).

14. Park, J., "MVP-Forth Expert System Toolkit," MVP-Forth Series Volume 4, Mountain View Press, Inc., (October 1983).

15. Reboh, R., "Extracting Useful Advice from Conflicting Expertise," Proceedings of the Eighth International Joint Conference on Artificial Intelligence, Karsruhe, West Germany, (August 1983), pp. 145-150.

16. Saaty, T. L., <u>The Analytic Heirarchy Process: Planning, Priority Setting, Resource Allocation</u>, McGraw-Hill International Book Company, London, (1980).

17. Shewchuk, J., "Justifying Flexible Automation," <u>American Machinist</u>, (October 1984), pp. 93-96.

18. Skinner, C. S., "The Strategic Management of Technology," <u>Autofact 4 Conference Proceedings</u>, Philadelphia, PA (November 30 - December 2, 1982).

19. Van Blois, John P., "Strategic Robot Justification: A Fresh Approach," <u>Robotics Today</u>, (April 1983), pp. 44-46.

20. Webster, R. and Miner, L., "Expert Systems: Programming Problem-Solving," <u>Technology</u>, Vol. 2, (January-February 1982), pp. 62-73.

Presented at the CASA/SME Flexible Manufacturing Systems Conference, March 1986

Artificial Intelligence and Distributed Processing: Key Technologies for the Next Generation FMS

by George M. Parker
General Dynamics

1. Introduction

Basic technologies involved in flexible manufacturing have been in use for more than 15 years. This is especially true in machining where the first flexible machining systems (FMSs) evolved from semiflexible transfer lines during the early 1970s. Many flexible machining systems still being installed today are using the same first generation FMS concepts, particularly in the realm of control software. However, currently emerging computer science technologies have opened doors that will enable flexible manufacturing to progress to the next control software generation.

The Advanced Machining System (AMS) program, sponsored jointly by the Air Force Wright Aeronautical Laboratories-Materials Laboratory (AFWAL-ML) and General Dynamics, is an effort to establish a fully integrated manufacturing system and demonstrate this system in aerospace machining. A key component of this program is implementation of an advanced flexible machining system capable of totally unmanned operation and completely integrated via two-way communication with General Dynamics' emerging computer-integrated manufacturing architecture. Early in the AMS development program (1982-1983) General Dynamics engineers analyzed the state of the art in FMS control software and found it lacking in key areas. Observations included the following:

o Full FMS integration into a company's material requirements planning (MRP), factory data collection, process planning, and CAD/CAM systems was not yet a reality. In all cases observed, a system manager or other persons were required to key in data resident in other computer systems within the company. Two-way communication between FMS and factory planning and control systems was almost nonexistent.

o At best, scheduling within the FMS was a semiautomated exercise in which the system manager was required to manipulate a simulation model or set priorities for jobs to be run in the system.

o In almost all cases, a central computer was used to control and schedule the system. Distributed processing advantages -- for example, allocation of different functions to different computers -- were only beginning to be explored.

o Local area networking technology was practically nonexistent. Most FMS control systems used simple point-to-point communications between the central computer and machine tools, guided vehicles, inspection machines, or other devices.

An FMS control system specification that addresses these perceived deficiencies was developed at General Dynamics in the wake of these observations. General Dynamics management recognized that no FMS control system existing then (in 1984) would completely satisfy the specification. Instead, efforts were focused on finding a vendor who could develop or integrate these capabilities into an FMS that would truly be a part of the next generation. The vendor ultimately selected was Westinghouse Industry Electronics Division. Personnel from Westinghouse, responsible for systems integration, teamed with DeVlieg Machine Tool Company personnel to submit the winning General Dynamics FMS proposal.

2. Overview of the General Dynamics FMS

The system should be considered from an overall perspective, as illustrated in the FMS physical model shown in Figure 1, before technologies are discussed that make up next generation control software. The portion of the system purchased from Westinghouse includes six five-axis DeVlieg 430R JIGMIL machining centers, each equipped with Automation Intelligence CNCs. Each machining center also will be equipped with tool interchange robots to automatically load and unload tools to and from the machine's tool magazine. Control Engineering, a subsidiary of Jervis Webb, is supplying the automated guided vehicles. Two LK coordinate measuring machines (CMMs) will be used for dimensional inspection. DeVlieg, in addition to supplying machine tools, will be responsible for mechanically integrating the entire system, including pallet changers, pallet storage area, and remote load/unload stands.

The basic system is nearing completion, and now efforts are being focused on meeting a further challenge -- a second phase system enhancement incorporating equipment not included in the basic procurement. This equipment also is shown in Figure 1. An Automated Storage and Retrieval System (AS/RS) is being procured from Munck Autech to provide automated backlog handling and transportation of parts from DeVlieg load/unload stands to a secondary processing area. General Dynamics also is procuring three robots, two of which will be used to automatically load and unload parts at the two DeVlieg load/unload stations. The third robot will be stationed at the other end of the AS/RS and will take machined parts and load them into an automated slurry deburring machine. After deburring is completed, the robot will remove the deburred part and place it in the final verification system. The final verification system will be used to check the part's surface finish, ensuring that the finish is free of burrs and other surface defects.

This additional equipment, which primarily was undefined when the purchase order was issued to Westinghouse, will create need for new control requirements in the future; therefore, the FMS controller must be easily expandable.

2.1 The Scheduling Expert System

As part of the AMS program computer-integrated manufacturing development, General Dynamics engineers are developing a sophisticated shop floor control system. This system, the lowest level of a full, closed-loop manufacturing resource planning (MRP II) system, will provide the FMS with all resources needed to perform its job (including all data resources). Figure 2 is an illustration of two-way communication between the FMS controller and the SFCS. General Dynamics' Shop Floor Control System (SFCS) also will perform macro-scheduling, i.e., it will assign a due date to each shop order to be manufactured in the FMS. Quantities and due dates of each FMS manufactured part will be passed to the FMS once each 24 hours. The FMS controller will be required to accept this order list and decide how to best fill orders without human intervention or input. General Dynamics' goal is to automate the system manager role so that this person will act as little more than a system watchdog.

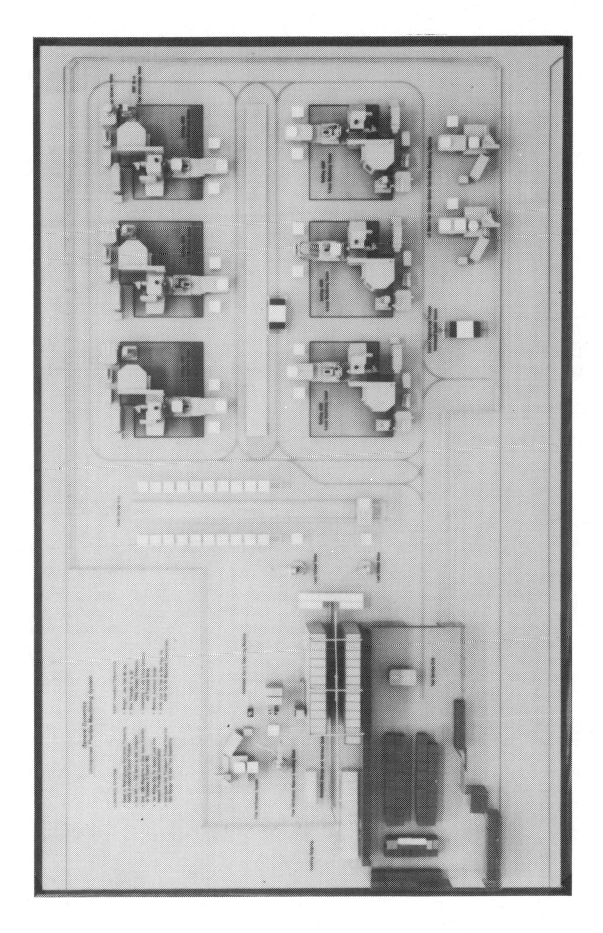

Figure 1 General Dynamics FMS Model

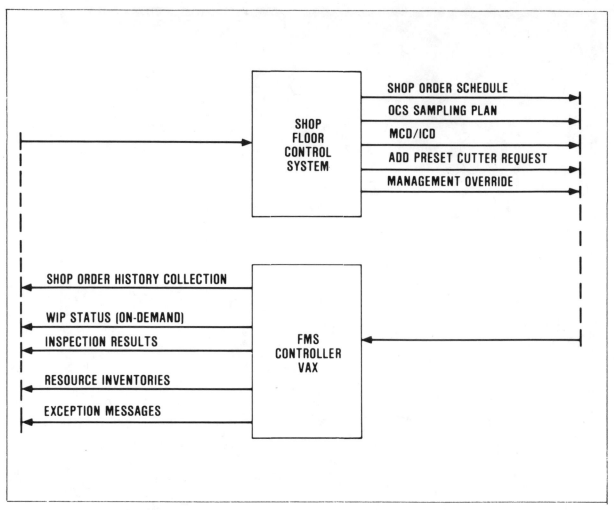

Figure 2 FMS Controller Communication With The SFCS

Scheduling shop orders to be processed in the FMS in order to meet due dates may sound like a trivial problem. However, FMS scheduling is expected to involve considerably more than meeting due dates. The scheduler also is expected to minimize part flow times, maximize use of primary equipment, and overcome adverse effects of bottlenecks, all while continuing to meet basic due dates. To make matters more complex, the scheduler is required to consider cutting tools while scheduling. Approximately 80 to 100 different types of parts can be made in this system, and only limited success has been achieved in reducing the number of different tools required to process these different parts. Considerably more than 200 types of cutting tools will be required to process the parts. The Westinghouse/DeVlieg system can automatically move tools around the system both to and from a central tool crib as well as between the machines. Tool carrier pallets transported by Webb guided vehicles and tool interchange robots located at each machine make this possible. The scheduler, therefore, is required to consider tool movements, in addition to part movements already considered.

Another factor that adds complexity to the scheduler is the process of loading parts (representing multiple part numbers) onto a single pallet. Fixturing cubes will be permanently mounted to each DeVlieg pallet, providing four vertical surfaces per pallet on which part holding fixtures will be mounted. Because of the low volume of aerospace manufacturing, construction of redundant fixtures for the same part is not cost-effective. Therefore, a single pallet in this system will usually hold holding fixtures for four different parts mounted on the fixturing cube. The scheduler must be intelligent enough to decide whether to load one, two, three, or more parts to fixtures on a given pallet. The scheduler will weigh efficiencies gained by combining orders for different parts and will compare extra flow time necessary to process multiple parts through the system on a single pallet.

Artificial intelligence (AI) was envisioned as the only workable basis for a scheduler that could handle requirements of both the initial system and the system yet to be defined. Development of such a scheduler by traditional methods would be difficult, primarily because the scheduling problem would be difficult to establish exactly enough to develop an algorithm to perform necessary or desired functions. Even if the problem were successfully defined and if an algorithm were successfully implemented, requirements of less well-defined Phase II enhancements would be difficult to incorporate into the algorithm. An artificial intelligence approach would enable Westinghouse to develop a basic scheduler capable of operation despite the constraints. The AI-based scheduler then could be modified easily by simply adding to the scheduler "knowledge base" and incorporating new features. The AI scheduler in fact is expected to become more intelligent as it is used in production, since it will encounter situations not envisioned during initial scheduler design and coding. As these new situations are encountered, new rules will be developed and added to the knowledge base to handle the new situations. Because of benefits AI offers, Westinghouse has been strongly encouraged to use the AI approach in FMS scheduler development. Westinghouse, in turn, has accepted the challenge and is implementing the AI scheduler for General Dynamics.

Even though artificial intelligence is a fairly new field, several expert system development systems exist to make AI development significantly easier. These development systems typically consist of the following:

o Inference engine - Software modules that act upon rules and facts contained in the knowledge base and that infer other facts and rules not explicitly stated. The inference engine is the heart of an AI expert system, since it actually derives solutions to types of problems under analysis.

o Knowledge representation language - A higher level language in which rules, facts, and other knowledge constructions can be written. The knowledge representation language is used by an applications engineer to actually build a specific knowledge base for types of problems to be solved by a system being developed.

o Compiler or interpreter - Used to map the knowledge representation language into computer-executable code. In most AI development systems the knowledge representation language is mapped into either Lisp or Prolog, the two most frequently used computer languages for artificial intelligence applications. The compiler or interpreter transforms the knowledge base written in the knowledge representation language into a form executable or usable by the inference engine.

o Development environment - User interface, including editors, debugging aids, and even graphic aids. This environment enables the application developer to modify, control, and monitor system behavior during development.

Westinghouse spent about two months evaluating various AI development systems before one was selected for use with the FMS scheduler. Westinghouse ultimately selected ART (the Advanced Reasoning Tool) as the FMS scheduler development system. ART is offered by Inference Corporation of Los Angeles, California. Through independent analysis at both General Dynamics and Westinghouse, ART was selected as the most powerful AI development system commercially available. An added benefit is that ART is available in a VAX version. Westinghouse previously had selected the VAX as the FMS scheduler host. Many other development systems only operate on specialized computers that have been adapted for Lisp. These specialized computers (for example, Symbolics and Lambda machines) are very powerful for stand-alone AI system development, but also are more difficult to integrate as part of the more complex FMS application. Use of a more standard general purpose computer was a necessity, a viewpoint shared at General Dynamics and Westinghouse and established through separate evaluations.

Once ART was selected and procured, Westinghouse personnel initiated attempts to implement the scheduler, a task that now is an ongoing effort. Since flexibility is one AI system attribute, the scheduler cannot be seen as a clearly defined computer program that is simply designed, coded, and debugged. The AI approach means scheduler implementation gradually will evolve rather than being developed to a final end point. Scheduler evolution will continue well past initial FMS implementation. After the FMS enters production, situations will be encountered that were not anticipated during initial scheduler design. The scheduler knowledge base will be easily modifiable to incorporate appropriate responses to new situations. Because of these modifications, the AI-based scheduler will become more intelligent as it is used in production. Therefore, presentation of scheduling algorithms used is simply not possible at this time. The scheduler Westinghouse is developing is much more flexible than conventional algorithms allow. The scheduler is an organic system whose makeup and actions change to adapt to new situations or to allow addition of new levels of sophistication.

However, as the AI scheduler is implemented, lessons being learned at both General Dynamics and Westinghouse can be discussed. ART, like many other AI development systems, is really designed as a stand-alone "world."

Interaction between ART and systems outside of ART is not achieved easily. Because of ART's large size and slow execution speed, implementation of the entire control system in ART is simply not feasible. Instead, Westinghouse has been forced to limit ART to the scheduling domain and to develop special interfaces linking ART and other FMS software operating on the VAX, which includes such vital entities as system and software databases that communicate with other portions of the FMS. Figure 3 illustrates additional software modules required to interface with ART. This interface to ART is extra software development that would not be required in a conventional algorithmic-based scheduling system.

A second lesson learned, a lesson Westinghouse prepared for well during basic system design, is that ART is very computationally intensive like all sophisticated AI development systems. Westinghouse learned early in the development process that ART will essentially dominate a VAX 11/750 processor so that nothing else will run on the same system. The problem is that ART software is very large, so large it will not fit into the VAX main memory even when expanded to 8 megabytes. Operating ART requires a very large number of disc accesses, so that portions of ART are continually swapped in and out of main memeory. This swapping slows the processor down to the point that no other program (other than the operating system) can be in memory with ART at the same time. The effect is impressive; even with scheduler running alone on a dedicated VAX 11/750, time required to compute a new schedule may exceed 10 to 15 minutes.

The solution to this dilemma is distributed processing. Distributed processing is use of separate but interconnected computer processors to perform different functions as part of an overall system. Distributed processing can be used to allocate each function to its own computer, sized appropriately for the function. An AI-based scheduling function might require the full power of a VAX super minicomputer, while another function, such as historical data storage, might be handled easily in a microprocessor-based system. Rather than reduce effectiveness of both functions by hosting them together on a single processor, considerable efficiency can be gained by letting each function run independently on different computers. In this manner, multiple processors can be used to work on different functions **at the same time** -- a feat that would be nearly impossible in a single computer without using a large and expensive mainframe rather than a VAX.

2.2 Distributed Processing

Like artificial intelligence, distributed processing is a technology that has been evolving in computer science for several years but so far has not been used extensively in discrete parts manufacturing. Westinghouse used distributed processing in controlling chemical and power plants before the decision to use this technology for General Dynamics' FMS. Use of a single "host" computer for FMS control presents other problems that Westinghouse could avoid through a distributed approach. Following are some of these problems:

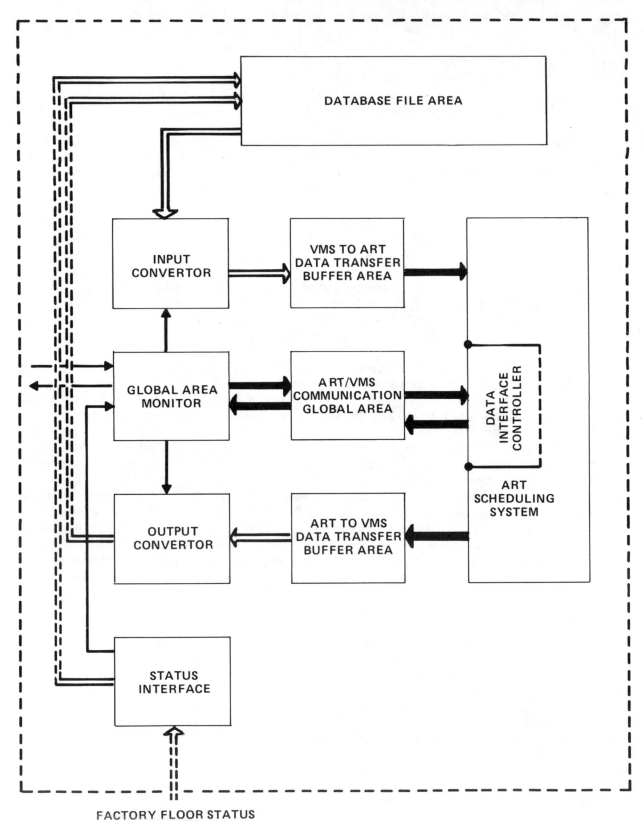

Figure 3 ART Interface to Other FMS Control Software

o Use of a single host computer means both real-time and background processes share computer resources. This sharing limits the sophistication of background processes (such as scheduling) since real-time control aspects must have top priority for access to computer resources (CPUs, disks, etc.). Achievement of total FMS scheduling automation would be nearly impossible in a single host system, because scheduling processes might cause severe degradation in real-time processes.

o Significant degradation also is possible when a crisis occurs that occupies the single host to the detriment of other processes being run on the same computer. Quick system control reaction to emergencies also will slow down responses to other FMS activities.

o A system using a single host computer is susceptible to downtime caused by host computer failure. In this situation, the entire system is forced to shut down because of a computer malfunction, even though other system components (e.g., machines, guided vehicles) remain operational.

To address these concerns, Westinghouse has proposed a new approach to FMS control. This new approach involves distribution of various control functions to individual computer processors. The best processor of appropriate size could be allocated to each job so that adequate computer power could be used everywhere in the system. Implementation of this philosophy enables a powerful VAX 11/750 processor to be dedicated for the most part to just one task -- scheduling. Without distribution, use of artificial intelligence would not be practical in the scheduling function.

Figure 4 is an overview of FMS controller architecture Westinghouse is developing for General Dynamics. This architecture is based on a series of products known as the Westinghouse Distributed Processing Family (WDPF). WDPF is the distributed processing system Westinghouse used previously for chemical process plant control and other continuous processes. WDPF is being modified to handle a more discrete part manufacturing environment -- the machining environment -- as part of the Westinghouse contract with General Dynamics.

A brief overview of distributed functions in General Dynamics' system will be given to promote better understanding of how distributed processing can be used for FMS control. First of all, note that the "glue" that binds various processors together in a distributed system is the local area network (LAN). In order for distributed processing to result in a single integrated control system, the network must be fast and reliable enough to ensure all processors receive required information from other processors as quickly as possible. In General Dynamics' FMS, the Westnet II local area network, also known as the WDPF Data Highway, will be used to integrate various processors. This network is fast, with a data transmission rate of 2 million bits per second. Even more important, this network uses a token passing scheme that ensures each processor on the network will get a chance to transmit data every tenth of a second, no matter how busy the overall network is.

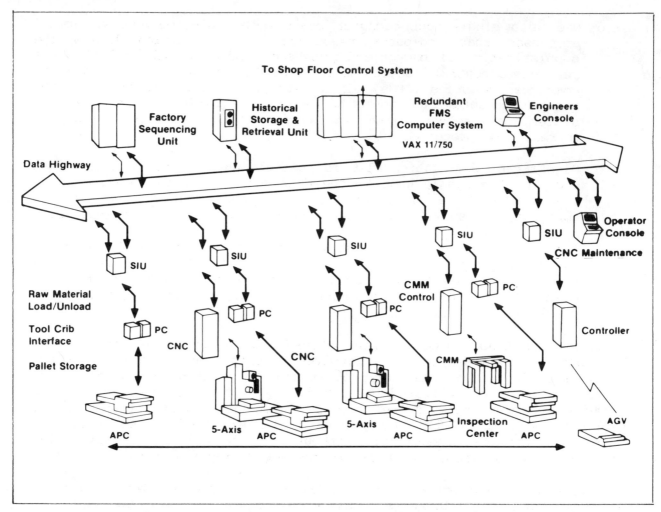

Figure 4   Westinghouse FMS Controller Architecture

Two VAX 11/750 processors will exist in General Dynamics' system. They will be closely coupled in a cluster arrangement so that if one processor fails the other can assume the failed processor's functions. This ensures that a failure in either VAX will not shut the system down. In normal operation, however, one VAX will serve as the primary processor and the other as backup. The primary VAX will host ART and the scheduling system; Westinghouse refers to this software as the master scheduling unit (MSU). The only other function for this VAX will be management of FMS databases. The VAX will therefore serve as the source for all information needed by other processors (including schedule data, NC part programs, inspection part programs, tooling data, etc.). This function is known as the File Server, since it is responsible for sending correct data files to any other processor that needs it. The File Server ultimately will get its information from the factorywide Shop Floor Control System through a separate communication link. This file server function, in conjunction with the master scheduling function, will essentially dominate the VAX's processing power so that all other FMS control functions must be distributed in order for the system to operate.

Real-time sequencing is another required FMS control system function. In the Westinghouse FMS, this function is distributed to two separate types of processors. Real-time sequencing of parts and resources between work-stations is performed by a separate computer known as the Factory Sequencing Unit (FSU). This separate computer is based on 16-bit microprocessor hardware. In fact, all other processors in the control system (other than the VAX cluster) will be based on 16-bit microprocessors. The FSU will handle its functions on the basis of schedules set within the MSU VAX. Since loss of the FSU would cause the entire system to go down, redundant FSU processors will exist in General Dynamics' FMS. The second FSU will automatically assume sequencing tasks in case of primary FSU failure.

Real-time sequencing within a single workstation (such as a machining center) is handled by another distributed processor, the Station Interface Unit (SIU). Figure 5 is an illustration of SIU architecture for a machining center. The SIU sequences pallet and tool movement within the machining center "cell," coordinating the machine tool, automatic pallet changer (APC), and tool interchange robot. This sequencing within the machining center workstation includes starting part program execution within the CNC. All SIU actions ultimately are determined by detailed schedule output of the MSU and are coordinated with other SIUs by the FSU. Note that each individual workstation (such as each machine tool) will have an individual SIU and that different types of SIUs exist for different types of workstations (such as CMMs, load/unload, tool crib, and even guided vehicles).

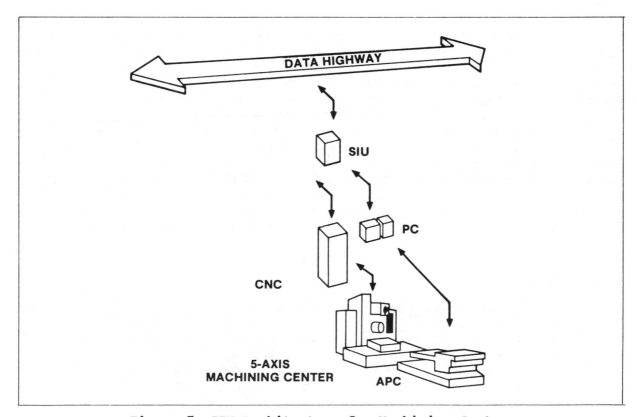

Figure 5 SIU Architecture for Machining Center

In addition to sequencing operations within a workstation, SIUs will handle communications with devices on the shop floor. Therefore, SIUs will handle protocol conversion, translating data transmitted over the WDPF data highway into a form understandable by device controllers such as CNCs, programmable controllers, CMM controllers, etc. Protocols used by the SIUs are programmable and can be customized to add different vendors' hardware to the WDPF network. SIUs will be used to add new equipment to the system originally purchased, such as the AS/RS, robots, and other devices. SIUs also will communicate with personnel in manned workstations, such as the tool crib, using a CRT.

The system is monitored in real time using another distributed processor, known as the Man-Machine Interface (MMI). MMIs include a 16-bit microprocessor (identical to the SIUs and FSUs), as well as a 19-inch color graphics screen and an associated keyboard. Figure 6 shows a bank of these MMIs. The MMI in the foreground, known as an engineer's console, includes a regular keyboard and a digitizer pad in addition to normal membrane style keyboards found on other MMIs. In General Dynamics' FMS, MMIs will be used for real-time monitoring of status data received from the data highway. Color graphics will be used to indicate the overall system, and more detailed screens will be called up to show status of a given machine or even a component of that machine (for example, the tool chain). The engineer's console will be used to create new graphics and displays for system enhancements. These MMI consoles will be used by the system manager to monitor the system and by maintenance to catch problems as fast as they occur.

The last type of distributed processor that makes up the General Dynamics FMS Control System is known as the Historical Storage and Retrieval System (HSR). Like most other computers in the control system, the HSR is based on a 16-bit microprocessor. But the HSR also includes a disc drive and nine-track tape drive, both essential to performance of system functions. The HSR monitors status and alarm data broadcast over the data highway and records this data at preset frequencies (once each second, once each 10 seconds, etc.). In addition to performing normal status data collection, the HSR also will log any system alarms automatically and will automatically record this data to magnetic tape at specified intervals. HSR's prime purpose is to provide a historical record of system performance for use in tracing and maintenance. This function is at a lower level of detail than monitoring and reporting functions handled by the File Server VAX.

3. Conclusion

General Dynamics' FMS control system as implemented by Westinghouse is potentially more powerful than a control system implemented in a single large host computer. Processors can work within their well-defined areas of responsibility simultaneously. Shared data can be simply passed along the network to other processors that require this data. This approach, based upon new concepts in computer hardware and communications software,

184

Figure 6   Man-Machine Interface

has given Westinghouse the computing power necessary to implement an artificially intelligent expert system to schedule the FMS. Artificial intelligence and distributed processing technologies are closely related in General Dynamics' FMS and both are necessary. The described system is being implemented now and soon will undergo initial acceptance testing. Full production implementation, including integration of AS/RS, robotic part loading/unloading, and automatic deburring and final inspection, is scheduled for the end of 1985.

The opinion of General Dynamics management, engineers, and various other technical personnel involved in FMS development is that General Dynamics' FMS is one of the first in a new manufacturing systems generation. This new generation will use the latest computer technologies to achieve total automation and full integration and will be completely driven by data fed down automatically from factory-level control systems. When this automation is achieved, the next generation FMS will be able to make intelligent decisions about how to schedule, route, and sequence the work load without human input. This FMS will use a combination of multiple computer processors to apply tremendous computing power to the shop floor. In this new FMS generation, the term computer-aided manufacturing will truly be replaced with the term computer-integrated manufacturing. The FMS Westinghouse is building for General Dynamics definitely represents one of the first steps into computer-integrated manufacturing and the new FMS generation.

CHAPTER 6

INSPECTION

Computer Vision Inspection from a CAD Database

by Steven B. Curtis
Digital Equipment Corporation

Truly automated assembly requires that all piece parts entering the manufacturing system be usable by the assembly equipment. Perhaps the only way to guarantee the quality of all parts is to inspect each critical dimension on every part. Automation of the inspection process could make the goal of 100% good parts possible.

While the perfect manufacturing system should produce perfect parts, there is a real need for automated inspection in the currently available manufacturing systems. An ideal manufacturing environment still requires inspection information, if even at the actual point of manufacture, to ensure the quality of its output.

Figure 1 is an overview of the automated inspection system. Note that the flexible inspection system conceptually consists of a measuring device attached to a manipulator. A computer vision system attached to a robot can therefore function as a flexible inspection system.

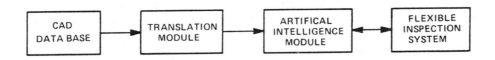

Overview of the Automated Inspection System

Figure 1

There are several unique elements of the prototype automated inspection system described below. Artificial intelligence routines simulate the way people manually inspect parts, eliminating the need for human intervention in the inspection process. To emulate a person, a multiaxis robot positions the computer vision system to view a small, predefined area. This significantly increases resolution and accuracy. The combination of artificial intelligence and a robot in an automated inspection system allows almost any dimension of a part to be inspected. The result of these features is a system that can automatically inspect parts.

THE NEEDS AND BENEFITS OF AN AUTOMATED INSPECTION SYSTEM

Much has been written about computers and their role in the manufacturing process. Often, though, simply adding a computer in hopes of solving a problem may create more problems than it solves. Expectations of a brand-new computerized system may be misdirected. This section addresses some of the indications for using an automated inspection system and some of the benefits of such a system.

Defective parts increase product cost. Faulty piece parts contribute to decreased yields, longer assembly times and more failures of the finished product. The cost of these bad parts can be enormous. These problems are aggravated further by automated assembly techniques. Robots generally do not deal well with defective parts. Much of the gain from automated assembly may be lost by more costly inspection procedures. In general, then, automated assembly requires that piece parts adhere more closely to their specifications.

Inspection costs decrease with automated inspection. Automated inspection results in less labor in the inspection process. This affects both the actual inspection process and the generation of the inspection procedure.

The automated inspection system can eliminate most of the manpower required to inspect a part. It should be able to run in a production environment with very little human intervention. Reasonably simple robots can manage material handling tasks, such as taking the parts from the parts carrier and presenting the individual parts to the inspection fixture. Thus, the inspection operation is potentially a lower cost procedure.

Generating inspection procedures for the automated inspection system requires very little additional effort once the part is designed. This additional effort goes into entering the tolerances of the objects to be measured into the CAD system. Once these tolerances are in the CAD data base the automated inspection system can, in principle, perform the complete inspection procedure.

Lower inspection costs have several benefits. Decreased inspection costs have several ramifications. The automated inspection system can inspect all parts in a lot, instead of inspecting a small sample of each lot, therefore guaranteeing that all parts reaching the production floor meet their specifications. Since the dimensions of all parts are stored, the automated inspection system could, as needed, statistically analyze the data. The analysis could forecast problems in the manufacturing process, thus eliminating the need to react to problems after they have occurred.

Quality of inspection improved. There are several other benefits that an automated inspection system would introduce. Eliminating people also eliminates the guess work and inconsistencies that they bring into the inspection process. Due to these inconsistencies, even 100% inspection by humans may not be very effective. The increased speed with which changes to the inspection process can be implemented allows changes in the design of the parts to be made very quickly. The general purpose nature of the automated inspection system requires fewer, if any, specially designed holding fixtures for the inspection operation. The automated inspection system could be located directly at the manufacturing process. This would allow immediate feedback about short term trends and reports about long term trends.

THE PROTOTYPE AUTOMATED INSPECTION SYSTEM

The Advanced Manufacturing Technology group at Digital Equipment Corporation in Colorado Springs has recognized many of the needs and benefits discussed above. A prototype automated inspection system is being developed using the guidelines set forth below.

Overview. Figure 2 presents a block diagram of the prototype automated inspection system. A production part is designed, complete with inspection information, on the CAD system. Once the part is approved, it is moved into the permanent CAD data base. The translation program translates the CAD information into a data base suitable for inspection purposes. This intermediate data base is referenced by the AI-1 module. AI-1 determines and simulates the inspection procedure and outputs general instructions to the AI-2 module. AI-2 performs the actual commands to operate the robot and vision system.

Figure 3 details most of the preliminary specifications for the prototype automated inspection system. Note that the characteristics of the prototype automated inspection system depend on specifications of either the robot or vision system. As more accurate robots and vision systems become available, the specifications of the automated inspection system will improve.

Basic requirement of a CAD system for automated inspection. Inspection information must be easy to enter into the CAD system and should modify the existing data base as little as possible. The inspection information should sit quietly on top of the finished CAD drawing and interfere as little as possible with the design process.

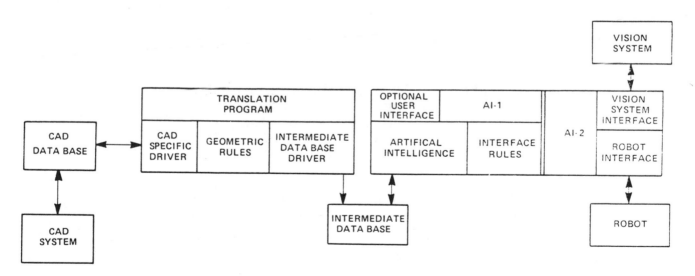

Details of the Prototype Automated Inspection System

Figure 2

Specifying features to inspect. The specification of features to be inspected can be a difficult problem. Few CAD systems provide tolerance information, and no CAD system provides inspection oriented entities. The automated inspection system must therefore add a method of specifying inspection data to the CAD design process. Three methods to specify this information were examined. One method is to add a comment that contains dimension, tolerance and reference information to each feature. The second approach considered used several different markers, one for each tolerance range, to indicate each inspection feature. However, both of these methods can be difficult to use and depending on the CAD system can be nearly impossible to implement.

The prototype automated inspection system uses the third approach to this problem. By placing a comment containing only the tolerance information near the dimension for a feature, the user can specify each feature to be inspected. In figure 4, the dimension entities were generated by the CAD system. The tolerance information has been added as notes. The dimension and tolerance information can be linked together, since the tolerance notes and the dimension entities are physically close together in the drawing.

Accuracy	Dependent on robot and vision system.
	Preliminary specifications:

Robot:
    Positional Accuracy: 0.002 inch

Vision System:
    Resolution: 256 by 256 pixels
    Field of View: 0.1 inch square

Overall Accuracy: 0.006 inch for
all supported   features

Supported Features   linear outside lengths
                      arc radius and location
                      hole location and radius
                      distance between hole centers
                      angle between two outside lengths

**Preliminary Specifications
of the
Prototype Automated Inspection System**

Figure 3

Translating the CAD data base. Every CAD system vendor has his own, proprietary method of storing the CAD model. The proprietary nature of the CAD data base can cause problems in writing the translation program. Fortunately, some CAD vendors provide routines to read data from their data bases. Most other vendors at least supply some documentation about their data structure. The proprietary data structures can affect the choice of the CAD system for an automated inspection system.

IGES may provide some assistance. IGES (Initial Graphics Exchange Specification) is a National Bureau of Standards specification. About thirty CAD vendors have adopted IGES. The specification defines a data base storage format that supports a reasonably complete set of CAD entities. A CAD vendor can support IGES by writing a computer program that translates his proprietary data base into the IGES data base. IGES is not meant to supersede each vendor's data base structure; it does allow different CAD systems to share and exchange CAD information.

The greatest benefit of IGES to a automated inspection system is the requirement to write only one translation program instead of a separate translation program for each of the many CAD systems available. This makes the use of an automated inspection system much easier in a multivendor CAD environment.

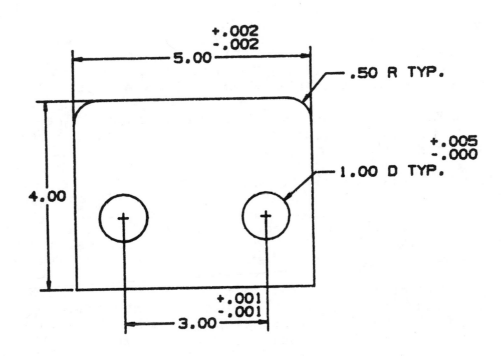

Example of Specifying Inspection Data

Figure 4

Translation program interfaces CAD and intermediate data bases. The translation program consists of three modules. The first module extracts the inspection data using routines that access the CAD data base. The second module determines the data to be extracted from the CAD data base. While all of the toleranced features are extracted, additional data about the geometry of the part is also needed. Artificial intelligence routines determine the necessary geometry of the part to be extracted. Due to the differences between CAD data base structures, the first two modules are quite dependent on the CAD system being used. The use of a different CAD system would require extensive reworking of these two modules. The third module writes the extracted data into the intermediate data base. All CAD data bases can use this module.

Intermediate data base holds all inspection data in a universal format. Each supported feature has a unique format in the intermediate data base. Feature specific information includes data about radii, length, angles, etc. Every entry in the data base contains information about the orientation, location, size and tolerance of the feature. The intermediate data base also stores the maximum and minimum dimension for each of the three axes.

Dimensions can be relative to other features on the part. The intermediate data base stores the location of each feature in absolute coordinates. By determining the location of each feature, the measurements are relative to actual position of each feature.

Two different implementations of the intermediate data base are being investigated for the prototype automated inspection system. The first method stores the data base as a disk file. This does not require that the CAD data base be on-line at all times, but does require an operator to rerun the translation program periodically. The second method recreates the intermediate data base for each inspection and therefore always reflects the current contents of the CAD data base.

AI-1 module emulates the human inspector. The AI-1 module uses artificial intelligence techniques to derive camera locations and views from the intermediate data base. It also ensures that the inspection system does not physically hit the part under inspection. A person can quite easily position a vision system and analyze its output. The exact algorithms that people use are not well understood, though. Therefore, a set of heuristic rules were derived by watching people position a vision system, and measure dimensions from the video output. The rules simulate the way people moved the vision system to inspect each feature.

Since the output of AI-1 is not specific to any one flexible inspection system, the AI-1 module is universal over all CAD systems and all flexible inspection systems. This is a result of the artificial intelligence routines used in AI-1. Both production and rule based implementations of AI-1 are being tested. There are a small number of problems with ambiguous geometry in wire frame models that require a person familiar with the parts to resolve the uncertainty.

AI-2 module handles automated inspection system specifics. The AI-2 module translates the the AI-1 output into instructions for the specific flexible inspection system. A small rule based artificial intelligence routine performs the actual translation. The general instructions from AI-1 translate quite easily into commands for the flexible inspection system used in the prototype automated inspection system.

The AI-2 module is very dependent on the flexible inspection system being used. A procedure for generating the rules used by AI-2 would decrease the amount of time required to write a new AI-2 module. It is also possible that many of the artificial intelligence rules could be encoded into a table. All that would then be required to generate a new AI-2 module is to put the instructions for the specific flexible inspection system into a table.

The flexible inspection system. The prototype automated inspection system will use a robot to move a vision system. The accuracy of a production quality automated inspection system should be on the order of one ten-thousandth of an inch. While the accuracy and resolution of the robot and vision system in the prototype will not be good enough for a production version of the automated inspection system, they will still allow a demonstration of the concepts discussed above.

By far the biggest requirement for both the vision system and the robot is the ability to understand externally defined points and locations. Many robots and vision sys ms have only a teach mode to enter commands and locations. This does not provide enough sophistication for a flexible system such as the automated inspection system.

CONCLUSION

A necessary and important step towards the factory of the future. Automated assembly is simplified if all parts are usable by the assembly system. Advanced inspection techniques can help in the production of 100% good parts. Automated inspection will move onto the production floor as the accuracies of both robots and vision systems improve.

Combining artificial intelligence with robots and computer vision systems creates not only an useful inspection tool, but can be applied to many areas of computer integrated manufacturing.

Presented at the CASA/SME AUTOFACT Europe Conference, September 1984

Automated Filmless Artillery Fuze Inspection

by J.A. Adams
A.P. Trippe
and
E.W. Ross
IRT Corporation

INTRODUCTION

Automated Digital Radiography is used by the U. S. Navy at Crane, Indiana to perform 100 percent inspection of fuzes used on Naval five-inch gun projectiles. This filmless x-ray system, called ARIES-100, was developed by IRT Corporation (San Diego, California) to rapidly inspect a variety of fuzes for critical defects. ARIES-100 can automatically inspect any one of eight different fuzes at a rate of 80 fuzes per hour. Additionally, the system can inspect the Navy Hi-Frag explosive loaded beaker for critical cavitation defects.

FUZE INSPECTION PROBLEM

In military terms, a fuze is that part of an artillery shell which triggers the explosion of the projectile. Fuzes, as shown in Figure 1, are usually electro-mechanical devices which assure that the ammunition is safe to transport, handle and load into the gun. After being fired, and at a safe distance from the muzzle, the fuze must "arm" itself and, when the target is reached, function and initiate the main charge. Fuzes are generally manufactured by modern automated machines, but due to the sheer numbers made, occasional defects can occur. If the defect is such that it causes the shell to prematurely detonate in the gun barrel, the result can be the loss of lives and a severe reduction in the total fire power for a fighting ship. For this reason, the Navy has classified fuze defects which can lead to an in-bore premature detonation as "critical". By this action, the Navy has specified that in these areas fuze quality be tightly controlled.

FILM CANNOT INSURE FUZE QUALITY

Since before World War II, fuzes have been nondestructively inspected for internal defects by x-ray film techniques. Film radiography produces a shadow image which shows the internal arrangement of components. Film has the inherent disadvantages that it must be developed before the image can be examined and once exposed, it cannot be reused. Because the cost of x-ray film fluctuates with the price of silver, many operations have reduced the number of radiographic inspections performed. This could open the door to a possible degradation of product quality. Film must also be read after it has been developed. Reader performance must be considered in evaluating the quality of the inspection results. Figure 2 illustrates how technology has progressed toward the reality of a cost-effective, reliable and fully automated radiography inspection capability. Technology first made man-in-the-loop systems available during the late 70's. High speed, solid state microprocessors made fully automated, radiographic imaging possible in the early 80's. Now in 1984, Automated Digital Radiography Systems are beginning to appear on the production line.

DIGITAL RADIOGRAPHY ADVANTAGES

Digital radiography was developed as a film alternative intended to eliminate the drawbacks of film. The advent of high-speed computers and modern video image

Figure 1. Typical U. S. Navy artillary fuzes

analysis concepts allowed x-ray fluoroscopy methods to be updated to a level which complements automated manufacturing and assembly operations. Digital Radiography is an implementation of computer aided testing technology.

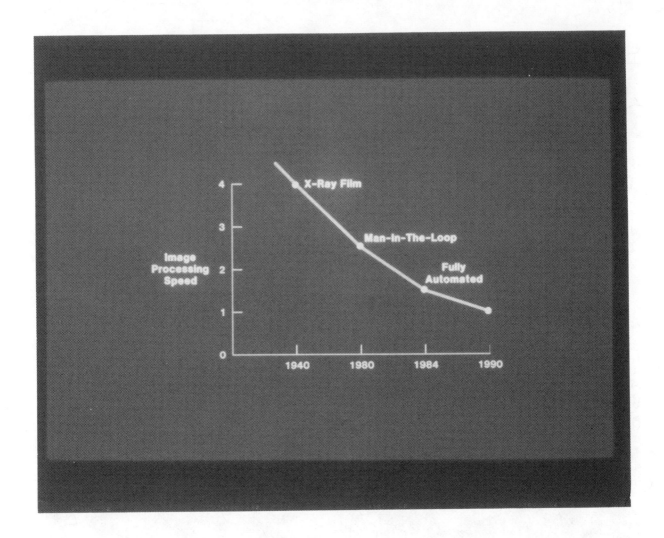

Figure 2. An Historical View of Radiographic Inspection Technology

The components of a digital radiography system, shown in Figure 3, consist of an x-ray source with collimator to direct the beam of radiation through the object under test, a fluoroscopic converter screen which produces the visible light image, a mirror that directs that image to a closed circuit low-light-level television camera which is located out of the x-ray beam path, an image digitizer which converts the analog TV signal into digital information, an image processor and associated high-speed computer for analysis, display, operator interaction, and image archieval storage, and a control subsystem which integrates the inspection operation with the plant assembly line. Enhanced, high-resolution images are generated in real time (or near-real time) by this digital radiography system, and the resulting images are analyzed for any critical defects that may exist in the item under inspection. Digital images are generated at standard video rates of 30 frames per second.

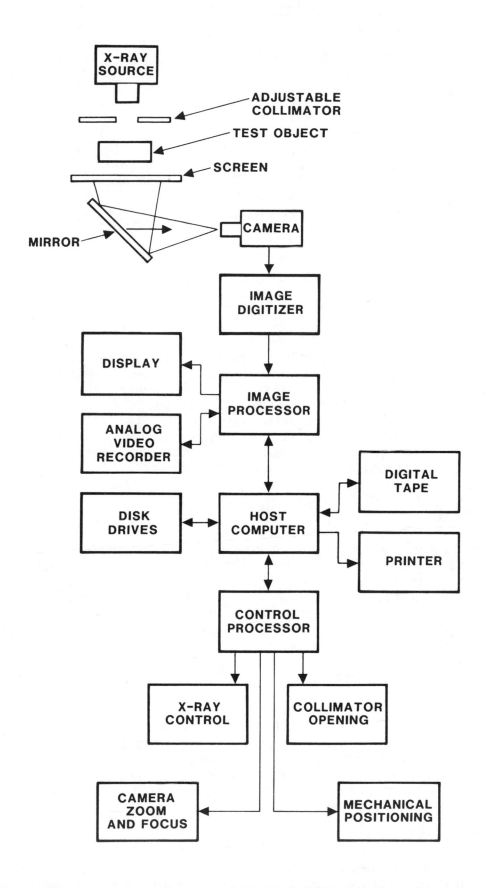

Figure 3. The components of an Automated Digital Radiographic Inspection System.

ARIES-100 DETAILS

With its history of working with manufacturing customers to solve production quality and control problems, IRT Corporation determined that in order to be of value, industrial radiography must provide information on a real-time or near-real-time basis, it must be compatible with production operations, it should eliminate sampling and allow for inspections to be conducted at production line speeds, it must be easily reprogrammed to accommodate new or changed product designs, and it must be user friendly so that production personnel can operate the equipment. Starting with these identified needs, IRT Corporation proposed to develop, fabricate and install an automated digital radiography inspection system for the inspection of critical defects in Navy fuzes. The concept is illustrated in Figure 4. The result of the Navy-sponsored project was the Automated Real Time Radiographic Image Evaluation System, referred to as "ARIES-100." With its 160 kV x-ray source located in a shielded cabinet and its computer controlled robotic loading, unloading and marking subsystems ARIES-100 has the capability to inspect eight Navy artillery fuzes for a variety of defects. ARIES-100 also inspects the high explosive "beaker" which is loaded into the two-piece "Hi-Frag" projectile.

MATERIAL HANDLING

The ARIES-100 automatically sequences the fuzes or beakers through a programmed inspection and marking (accept or reject) sequence. As shown in Figure 4, a robot is used to accomplish loading operations. Other systems designed by IRT have employed part feeding mechanisms based on a rotating index table or a linear belt feed rather than the robot arm design. Correct rotational and vertical orientation of the fuze is attained under automatic computer control. Two views of each fuze are used to clearly image the important details. Interlocked and explosion proof electronics also provide a high level of radiation protection and personnel safety in the presence of explosive-loaded ordnance items.

IMAGE GENERATION

The radiation source is a commercial 160 KV x-ray tube housed in a shielded cabinet. After penetrating the fuze, the beam of x-rays impacts the fluorescent screen, which generates a visible light shadow image of the internal components. The composition and thickness of the screen affects the brightness and resolution of the image. A high-quality optical mirror allows the camera to view the screen without being placed in the direct x-ray beam. Irradiation of the camera can cause damage to the lens and electronics. Specific radiation shielding around the camera provides an additional level of protection. Retraction of the mirror allows the camera to view a luminescent test panel, providing a means for focusing the camera lens.

In the ARIES-100 video chain, a low-light-level isocon closed-circuit television camera converts the image to an electronic signal in standard video format. Other applications may use a vidicon, SIT, CID or CCD camera type. Image zoom capabilities are obtained by moving the camera toward the screen and refocusing the lens. Remotely controlled motorized camera positioning and lens focusing allow a continuous magnitification capability.

IMAGE PROCESSING HARDWARE

The image processor subsystem, shown in Figure 5, accepts the video signal from the camera and performs the image digitization, display, enhancement, analysis

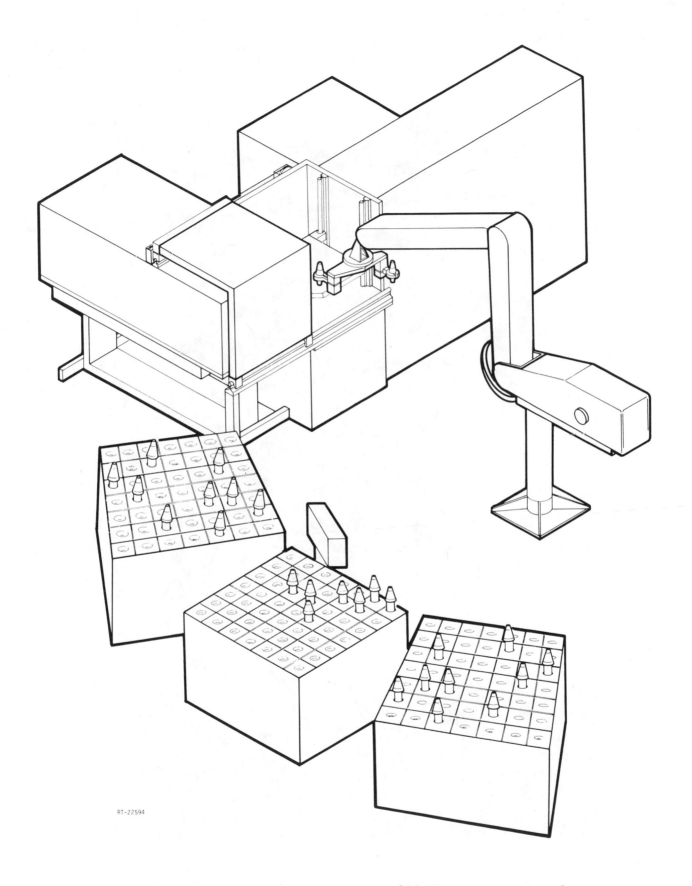

RT-22594

Figure 4. A concept drawing for the ARIES-100 Fuze Inspection System

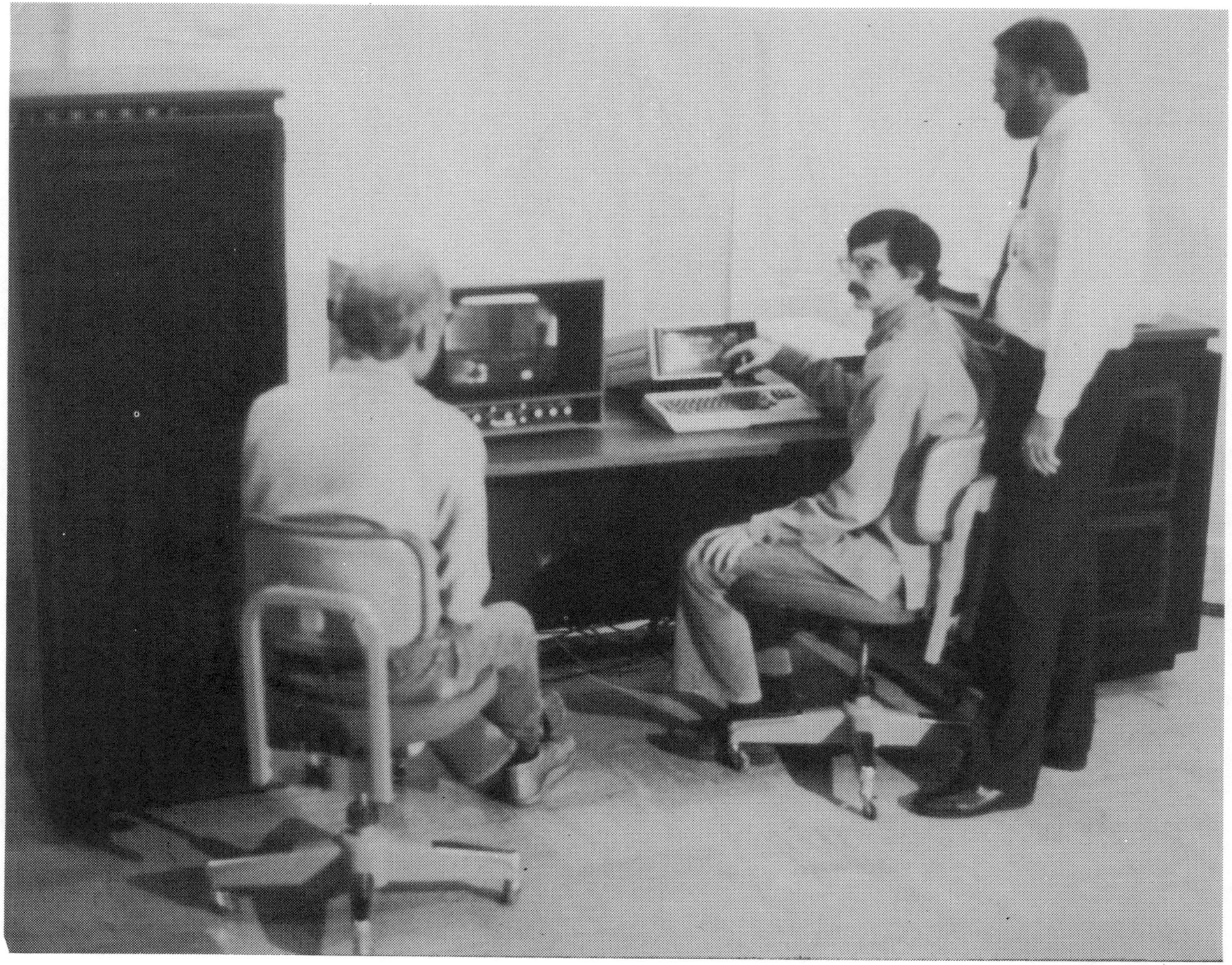

Figure 5. The ARIES-100 Image Processor Subsystem

and storage functions needed to achieve completely automated control of the inspection operations. The subsystem consists of state-of-the-art electronics which digitize, process and store picture arrays which consist of 512 x 480 x 8-bit elements, each representing a brightness level at a specific location in the image. Each digital image therefore consists of 245,760 picture elements or pixels which range over 256 gray levels from white to black. For other applications, 1024 x 1024 arrays of 16-bit pixels are available. The camera generates a new image 30 times per second. The image is processed during the 1/30th of a second available for each image frame. The recent availability of high-speed electronic devices has led to the development of digital imaging systems which have the ability to sum, smooth and enhance the radiographic image at those production line speeds commonly encountered in industry. The digital image can be displayed on black and white or color monitors, reproduced on a hardcopy image printer, stored in analog format on videotape cassettes, stored in digital format on a streaming digital tape, or placed in computer memory where it can be analyzed by automatic defect recognition software. The operator can command, control and communicate with the image processor through a keyboard and a touch panel, shown in Figure 6. The touch panel is a pressure-sensitive display where the desired operation is selected from a menu of options. This touch panel unit eliminates the need for a large operator console with an assortment of knobs, buttons and switches.

IMAGE PROCESSING FUNCTIONS

The image processing hardware subsystem is integrated with firmware enhancement functions, a few of which are noted below:

1. Grab takes a single frame of video information and stores the result in computer memory.

2. Frame average integrates consecutive images on a pixel-by-pixel basis for as many as 256 frames. This process minimizes random noise and enhances the contrast of true features.

3. Image subtraction displays those pixels which vary from one image to another. An image may be subtracted from a standard image to identify defective assemblies or to detect changes in a dynamic scenario. This technique is the basis for many effective pattern matching procedures used for continuous, on line inspection operations.

4. Gray-level transformation is a class of functions which enhance an image by manipulating the gray levels based upon a reassignment table or procedure. In its simplest form, the range of brightness levels between black and white is compressed so that the 256 gray levels span a smaller scale. The pixels contianing the feature data are "stretched" over the entire range of 256 gray levels. This function is useful for pulling small contrast features out of an image with little overall contrast. The use of a lookup table to assign specific input pixel values to designated gray levels produces a custom transformation which can be useful for highlighting features of a single component in the image.

5. Measurement of linear distances across the image is accomplished by using a cursor to mark two points on the image. System software will then calculate and display the distance between the points. The cursor may also be used to designate a window on the image. The area within the window will then be generated and displayed by the system software.

Figure 6. The ARIES-100 Touch Panel and Keyboard

6. The distribution of the gray levels within an image can contain important information. A histogram computation function can display the distribution of gray levels for an entire image or within a window. This information is often useful for determining the presence of small component parts. The number of pixels for a specific gray level can identify a broken or damaged part.

7. Other gray-level procedures include plots of the gray level values on a horizontal or vertical line through the image, and computation and display of the ratio of two gray levels within the image or a window. Component detection within an assembly is often accomplished through use of these techniques.

8. Selective density slicing is a process which allows for the display of only selected gray levels within an image. All gray levels not selected are displayed as white or black.

9. Zooming is accomplished by expanding each pixel of the image. In this manner, a portion of the image can be expanded to fill the entire 512 x 512 display. This technique does not enhance resolution, but it is useful for studies of detail in a selected area. Pan and scroll features can be used to roam over the image in this zoomed mode.

10. Date, lot number, inspector's name and other production-related information is also permanently stored on the digital tape or video cassette images when typed in from the keyboard.

AUTOMATED IMAGE PROCESSING

The ARIES-100 utilizes automatic defect recognition software to integrate the entire inspection process. The computer controls the loading, unloading and marking of the fuzes. It directs the image acquisition and enhancement procedures. It also uses the image processing functions to methodically detect irregularities, improper assemblies, missing parts, and most importantly, armed fuzes. ARIES-100 makes a final determination as to whether the fuze under test meets the military specification or not. In examining a fuze, the ARIES-100 first adjusts the picture for gain and offset, analyzes the rotational position of the fuze, moves the fuze to the first viewing position, acquires a picture, centers the view, rotates to the second view and acquires the second picture and merges the important part of the second view with the important part of the first view, to form the composite picture which the computer will analyze. The computer-based image processor then performs the necessary data processing and analysis to understand the picture and make a judgement upon the quality of the fuze. Figures 7 and 8 illustrate a typical automated defect inspection sequence involving a setback pin. The exact routines that perform the determination of accepting or rejecting the fuze are proprietary. The false accept rate is calculated to be one in one million, and the false reject rate is four in one thousand.

In those cases where the detection of defects is extremely critical, or where human reasoning must be applied, the automatic defect recognition software can be configured so that the operator is informed by the computer than a potential defect has been identified. After examining the suspected areas, the operator can make the final determination of quality based upon visual examination of the enhanced image presented on the display. This action activates the system to perform the proper accept or reject operations.

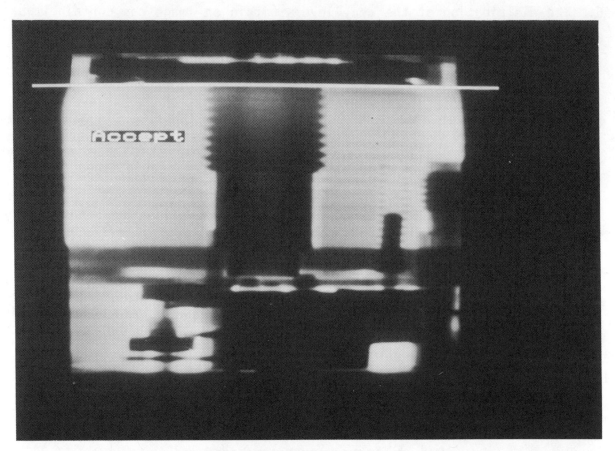

Figure 7.   An ARIES-100 Image of an Acceptable Fuze

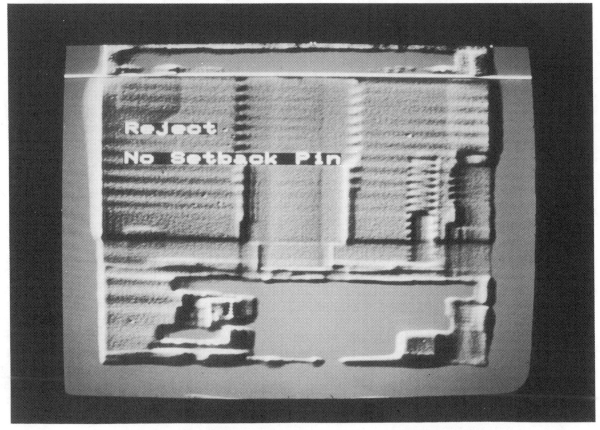

Figure 8.   An ARIES-100 Image of a rejected fuze

"TEACH ME" FEATURES

In order to provide flexibility for the Navy to expand the use of ARIES-100, IRT developed the teach mode (TEACH-ME tm) of operation. In this mode, the computer stores a sequence of inspection operations performed manually by an operator working to acquire, enhance and analyze images from new fuzes, or for fuzes which have had engineering changes incorporated into their design. When the operator is satisfied with the sequence of steps, he commands the system to save the sequence which then can be repeated over and over for automated testing of additional fuzes. In this manner, new inspection sequences can be developed without the assitance of a trained computer programmer.

CONCLUSION

When reliability is critical for raw materials, in-process components or assembled end items, quality can be assured using an ARIES digital imaging system. For the inspection of fuzes and other high reliability ordnance, it is now possible to configure a high technology solution for your application. Examinations are accomplished without the cost of film, and all testing is conducted on a real-time basis. Critical accept/reject decisions which cannot tolerate human error can be consistently performed by a fast, on-line and fully automated ARIES digital imaging system.

CHAPTER 7

PROCESS CONTROL

Reprinted from *Mechanical Engineering*, October 1985

ARTIFICIAL INTELLIGENCE IN PROCESS CONTROL

To apply expert systems to process control, one can begin by asking, "How would my best operator handle this problem?"

KENNETH W. GOFF
L&N INSTRUMENTS UNIT
LEEDS & NORTHRUP COMPANY
NORTH WALES, PENNSYLVANIA

Although artificial intelligence has been developing steadily since the 1950s, its capacities in such areas as voice and visual recognition still fall far short of human behavior. Early efforts involving search techniques and computational logic achieved limited levels of success with relatively elementary problems and games. However, because these approaches require a search space that expands exponentially with the number of parameters involved, they are not practical for larger, real-life problems. In such applications as process control, expert systems offer the possibility of circumventing these limitations, which explains why the growth potential of this segment of the AI market has been projected at over one billion dollars by 1990 (Figure 1).

STRUCTURE OF EXPERT SYSTEMS

Conventional systems are characterized by hard data and algorithms. They do not cope well with uncertainty and the reasoning supporting their decisions is not evident in the programs. An expert system, on the other hand, is built on a base of knowledge patterned after that of a human being, and it can solve problems using only incomplete and uncertain information. It is also distinguished by the ability to trace the path of reasoning that was followed in reaching a particular conclusion.

The structure of an expert system is shown in Figure 2. The knowledge base consists of general information about the problem in the form of *knowledge rules,* together with *inference rules* that control the way conclusions are reached. Knowledge rules are often based on if-then statements such as those in the following simple example: *If* the power supply fails *and* backup power is available *and* the reason for failure no longer exists, *then* switch to the backup power supply.

Each of these statements represents one basic chunk of knowledge, which can be added to or deleted from the knowledge base. Because the statements are not bound up with data and program instructions, as in conventional machines, they can be used to explain how decisions were reached.

The control structure in Figure 2 includes the *inference engine,* which controls and organizes the steps leading to the solution of a problem. It receives information from the data base, which has first been supplied with all the necessary input. It should be noted that the data and program instructions are not combined in one memory, as in the conventional von Neumann computer architecture. Conclusions are reached by means of a chain of if-then rules, forming a line of reasoning. A *natural-language interface* communicates the results to the user, including explanations of the reasoning process.

REQUIREMENTS AND LIMITATIONS

Building a knowledge base for a demanding application is a major effort that requires intensive interviewing of at least one expert by a knowledge engineer. Typically, an hour of effort is required for each chunk of knowledge, and tens of thousands of chunks may be needed to define the knowledge base of a large application. The job may take a team of two to five skilled engineers (who are in short supply) five or more man-years to complete.

The processing of knowledge rules and inference rules requires a programming language capable of handling symbols and lists rather than arrays of numbers. The two most widely used languages for programming expert systems are Lisp (short for List Processing) and Prolog (Programming in Logic). In addition, natu-

Segment	1983	1987	1990	Growth Rate
	(Millions of dollars)			(percent)
Expert Systems	16	800	1243	86
Natural-language Software	18	190	1090	79
Computer-aided Instruction	7	30	100	46
Visual Recognition	30	230	860	60
Voice Recognition	10	50	230	56
AI Hardware	25	200	500	54
Total	106	1500	4023	68

Source: Sperry Corp.
Engineering Manager—January 1985

Figure 1. Artificial intelligence market.

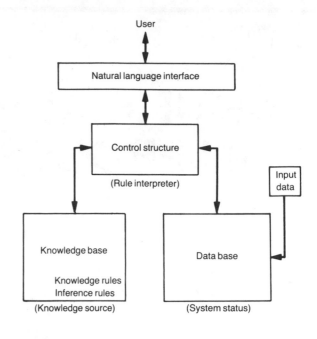

Figure 2. Basic structure of an expert system.

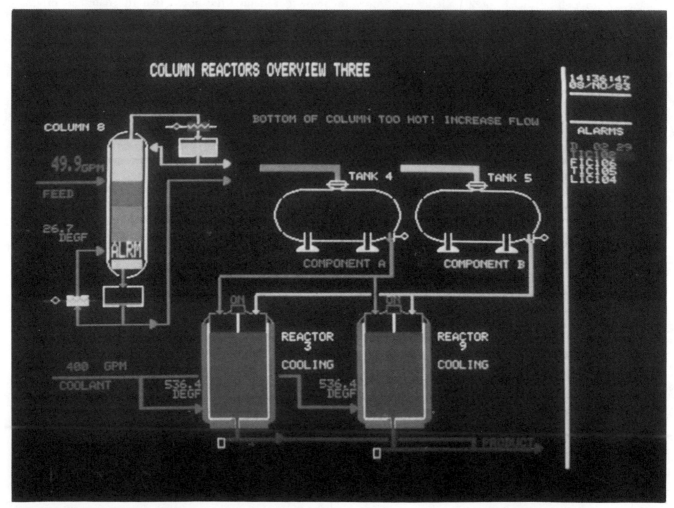

Figure 3. Alarm operator guide.

ral-language processors are expert systems in their own right that are based on the English language.

To handle the programming languages efficiently, powerful computers with very large memories are necessary. Special Lisp processors are now available for these applications, and parallel architectures are increasingly providing greater processing power. With the rapidly declining cost of processors brought about by VLSI technology, it is becoming feasible to use hundreds of these small processors together, thus approximating the parallel organization of the human brain. For example, an optical crossbar switch is being developed to provide communication among processors in a manner similar to a telephone exhange, but at data rates exceeding 100 megabits per second.

CONTROL SYSTEMS

The design of effective process control systems has for many years been approached with the question, "How would my best operator handle this problem?" Applying expert systems to process control systems involves much of the same common sense. But knowledge engineering and its associated tools, hardware, and software offer the opportunity of consolidating past gains and providing more powerful solutions to many problems. Alarms and control systems of a wide range of sophistication are among the areas that have benefitted from the application of expert systems.

When an operator is confronted with one or more process alarms, he must determine as quickly as possible the causes of the alarms and the best course of action. But sometimes the situation will call for additional information, the advice of experts, and time to evaluate alternative strategies. For example, many alarms are set off by invalid inputs such as bad sensors or equipment that has been shut down. To avoid an improper response, all the causes must be sorted out and corrected. An expert system can provide the operator with the necessary expertise and judgment and execute its decision with a computer's speed and precision.

Simple Operator Guide. An operator guidance system such as the one illustrated in Figure 3 is effective when the relationship between the alarm and the required action is rela-tively straightforward. This CRT display from a Leeds & Northrup Max 1 distributed system identifies the alarm with flashing red messages both in the process vessel itself (on the left of the screen) and in the alarm tabulation column at the right. A message: BOTTOM OF COLUMN TOO HOT! INCREASE FLOW also appears.

Chlorine Vaporizer. Another example that involves only a few variables but is more complex is the chlorine vaporizer process [1] illustrated in Figure 4. Liquid chlorine enters the vessel from the right and is heated, producing chlorine vapor which leaves the vessel at the top. The liquid level in the vessel is controlled by adjusting the flow of steam until the chlorine is vaporized at the same rate at which it enters the vessel.

An erroneous low level signal to the level controller could reduce the steam flow to the point that the vessel could fill with liquid chlorine. This situation could be hazardous if it resulted in the vapor line filling with liquid chlorine. The available data could be analyzed to detect the erroneous signal and appropriately warn the operator with an alarm rule such as the following:

If the steam mass flow is less than the steam-to-chlorine ratio times the chlorine vapor mass flow,
and the output of the level controller is greater than 80 percent,
and the level LOW-DEVIATION alarm is on,
and the level is not increasing,
and the chlorine feed mass flow is greater than the chlorine vapor mass flow,
and the standard deviation of the vapor mass flow is less than 5 percent,
then send this message: CHLORINE LEVEL SIGNAL TO CONTROLLER IS PROBABLY ERRONEOUS. CLOSE CHLORINE FEED VALVE IMMEDIATELY.

The reasoning behind this conclusion could be supplied to the operator in a comprehensible form, substantially increasing the credibility of the alarm message.

This example illustrates several important points in the application of expert systems to process alarms. First, the situation may demand knowledge that is well beyond the operator's ability. Second, correct analysis of the available information is particularly crucial in cases where treating a symptom rather than the cause of a problem could increase the risk of accident. Finally, when the recommended action contradicts what is indicated by the instrumentation, it is desirable that the operator be provided with an explanation that will help him understand and accept the recommendation.

Alarm Advisor. For applications of a larger scale, such as refineries, an expert system can process alarms by interfacing with a plant's conventional distributed control system [2]. In one system, knowledge-base inferences are processed by a special Lisp microprocessor while calculations and data checks are performed in parallel on a conventional processor. Ethernet bus and process interface modules can be added to bring other necessary data into the system. The results of the alarm analysis are displayed on a CRT.

Alarm Filtering. In electric power systems, the extremely high volumes of data and the necessity of responding quickly to problems make the management of alarms a particularly difficult task. Systems may now handle over 1000 alarms per hour and up to twice that many in a critical five-minute period. A recent study of alarm processing by the Electric Power Research Institute recommended that operators be supplied with transformed data rather than with volumes of raw information. These should provide a better picture of the entire power system and of the conditions causing the alarm and a comprehensive basis for decision making. Alarm priority should change dynamically with system conditions.

An EPRI study of an intelligent alarm processor that uses an expert system to meet those objectives produced encouraging results [3]. The processor was programmed in Lisp and tested on a training simulator. In a simple case involving two sets of breaker openings, four alarms were displayed to the operator out of the 21 that resulted. In a more complex case, 14 alarms were reduced to nine, and in the most complex test, 124 alarms were reduced to 30 plus two special messages. Although these were only preliminary tests, they appear to demonstrate the feasibility of applying expert systems concepts to these important alarm systems.

Expert systems have been applied at all levels of process control, from

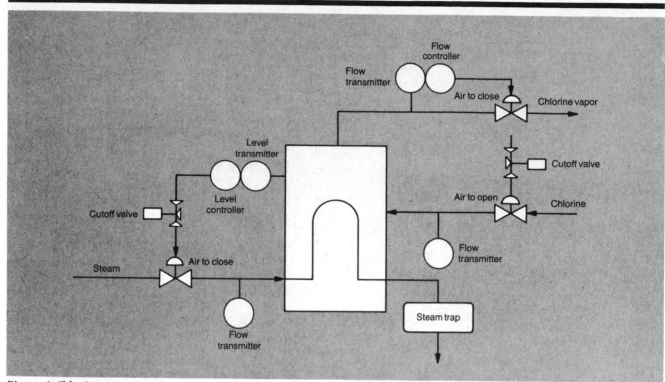

Figure 4. Chlorine vaporizer process control.

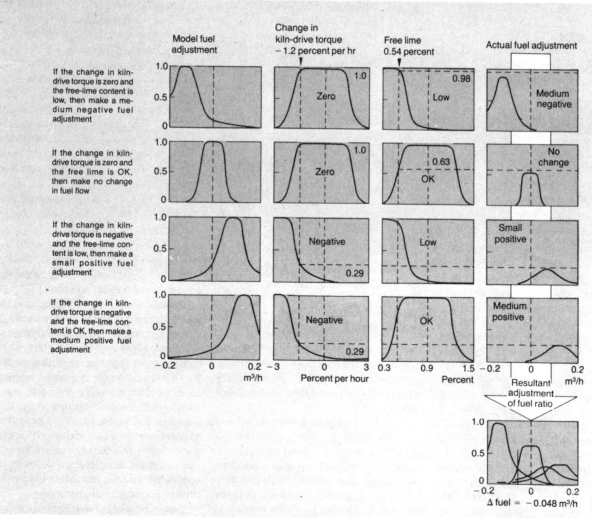

Figure 5. Fuzzy-logic process controller for cement plant fuel control.

single-loop controllers to large integrated systems. Some examples spanning this range follow.

Self-tuning PID Controller. A controller can use pattern recognition to identify step-like disturbance responses in the loop-error signal [4,5]. These are then used for tuning the controller. Initial tuning values are provided or they are generated by a pre-tune mode that applies a step disturbance to the process. The operator must specify a noise-threshold level and the maximum waiting time for the second peak of a step-like disturbance response. This, of course, is based on the process dynamics.

When the loop error exceeds twice the specified noise-threshold level, it is tested against a series of rules to determine whether it qualifies as the first peak of a step-like disturbance response. Similarly, the controller searches for the second and third peaks to define the classic oscillatory error response for a step disturbance. The period, overshoot, and damping calculated from the three peaks characterize the error response, and this information is used for a conventional Ziegler-Nichols tuning calculation. Confidence limits for each calculation are determined by the size of the error peaks compared with the noise threshold and by how well they match a typical damped sine-wave profile. The controller implementation includes several hundred rules that are expressed as if-then statements and nested as many as five deep.

Cement Plant Control. To implement a control system based on an expert system, a way must be found to translate the expert's qualitative rules of thumb into terms that the computer can use. The concept of fuzzy sets can be useful here [6].

An extension of the conventional mathematical concept of sets, fuzzy sets provide a gradual transition from membership to nonmembership in a particular set. The transition is defined by grades of membership from 0 to 1. For example, if Andy is a 48-year-old member of a technical society, to what degree does the label, "young member" apply to Andy? An answer of 0.3 indicates a grade of membership that recognizes the gradual transition from membership to nonmembership in the set of young members.

In cement plant control the plant operator monitors many variables, including temperature, gas composition, and speed, and applies to them perhaps 40 to 50 rules such as the following: If the oxygen percentage is *rather high* and the free-lime and kiln-drive torque rates are *normal, decrease* the air flow and *slightly reduce* the fuel rate. To apply the expert rules employed by the plant operator, the control system designer must relate qualitative terms such as "rather high" to measured quantities. This can be done by graphically relating the measured quantities to degree of membership in the fuzzy sets labeled "rather high" and "normal." Similarly, control adjustments can be related to those qualitative terms.

Figure 5 illustrates the application of this approach to determining fuel adjustment in a cement plant. Four rules of thumb are cited in the left-hand column. The next column depicts a model fuel adjustment pattern associated with each rule of thumb. The next two columns define the fuzzy sets associated with kiln-drive torque and free-lime content, and the pointer indicates the current value of each variable. The desired fuel adjustment resulting from the application of each rule of thumb to the current kiln conditions is shown in the right-hand column, and the weighted summation of these produces the actual fuel adjustment shown in the lower right corner.

Although it may appear somewhat complex, this approach to process control results in a program that is relatively simple to adjust by adding rules or changing the fuzzy logic. It is also easy to communicate the rules to operators. The system, which was first applied to cement plant control in Europe in 1980, is now used in several plants.

Power System Restoration. When one or more breakers open during an emergency, causing a brownout or a blackout, it is necessary to restore the power system with a maximum of speed and safety. Because the system operator's rules of thumb appear to play a key role in such situations, an expert system can facilitate successful restoration.

In one system, a top-down, or goal-oriented, approach is used [7]. Each of the failed buses is evaluated against two restoration patterns: energizing with a line that had fed the bus before the fault, or energizing with a standby line that had not fed the bus before the fault. If either pattern applies, power can be restored. The system is then checked for overload with a load-flow computation. Sixteen production rules such as the following are used in the analysis:

● If the component is not charged and it is not faulted, then it is concluded that the component must be restored;

● If the source is charged and its restoration pattern number is either 1 or 2, then it is concluded that the source is restorable.

The restoration plan generated by this approach will separate the blackout area into two parts and reconnect them successfully.

Because the computer program cannot cope with inconsistencies, the system designer must assure the consistency of the knowledge base. This problem, which is common in expert systems, has led to the concept of a *knowledge czar* responsible for the internal consistency of the rules in the knowledge base. The speed of the process control system is another problem. The Lisp program used in the power system restoration runs rather slowly on a DEC system 20/20; a Lisp-oriented processor is needed to run the program at speeds consistent with on-line operation. ▣

REFERENCES

1. Fortin, David A., *et al.,* "Of Christmas Trees and Sweaty Palms," *P.oceedings,* Ninth Annual Advanced Control Conference, West Lafayette, Indiana, September 19–21, 1983, pp. 49–54.
2. Moore, Robert L., *et al.,* "Expert Systems Applications in Industry," *Advances in Instrumentation,* Vol. 39, Pt. 2, ISA, 1984, pp. 1081–1094.
3. Wollenberg, Bruce F., "Feasibility Study for an Energy Management System Intelligent Alarm Processor," *Proceedings,* IEEE Fourteenth Power Industry Computer Applications Conference, May 6–10, 1985.
4. Kraus, Thomas, "Self-tuning Control: An Expert Systems Approach," *Advances in Instrumentation,* Vol. 39, Pt. 2, ISA, 1984, pp. 695–704.
5. Kraus, T.W. and Myron, T.J., "Self-tuning PID Controller Uses Pattern Recognition Approach," *Control Engineering,* Vol. 31, No. 6, June, 1984.
6. Zadeh, Lofti A., "Making Computers Think Like People," *Spectrum,* Vol. 21, No. 8, August, 1984, pp. 26–32.
7. Sakaguchi, T. and Matsumoto, K., "Development of a Knowledge-based System for Power System Restoration," *IEEE Transactions on Power Apparatus and Systems,* Vol. PAS-102, No. 2, February, 1983, pp. 320–329.

Reprinted from *I&CS—The Industrial and Process Control Magazine*, March 1985

Artificial intelligence moves into industrial and process control

No longer confined to the research laboratory, AI will soon become a valuable tool on the factory floor. Here's a look at the issues, applications, and tools available.

Dr. Richard A. Herrod
Senior Member of Technical Staff
Texas Instruments Inc.
Dallas, TX 75265

Barbara Papas
Strategic Analyst
Texas Instruments Inc.
Dallas, TX 75265

Artificial Intelligence (AI) is becoming an important factor in shaping the future applications of computers to many disciplines, including manufacturing and process control. Speech synthesis and recognition, vision, natural-language programs, and Expert System technologies are now being applied in a wide range of activities. These include molecular engineering, geophysical analysis, design and specification of sophisticated circuits, and, of course, industrial control applications.

Although AI promises many great things for the future, it is not a mature technology. But as the technology develops and costs decline, it should spread rapidly in a market that is even now eager for its benefits. The result will be an increase in individual and social potential stemming from the preservation and distribution of knowledge, and the release of workers from boring tasks that inhibit productivity. The question is no longer whether AI will have an impact, but rather how much and how soon.

The early days

AI research has been going on since the 1950s (Fig. 1). The work, carried out mostly in university research laboratories, had been aimed primarily at developing computers that can think and learn, much in the way that people do. The research was so challenging that little effort was expended dealing with "real" problems; that is, the needs of industry. Instead, researchers explored basic principles by working on theoretical problems and games, such as teaching a computer how to play chess.

Several years ago, AI moved out of the lab and into the commercial marketplace. The driving force for that move was the development of successful Expert Systems and the increasing availability of reasonably-priced computers that can deal with the large and complex programs characteristic of AI.

What it's all about

Basically, AI is a branch of computer science that addresses problems requiring human-like reasoning and in-

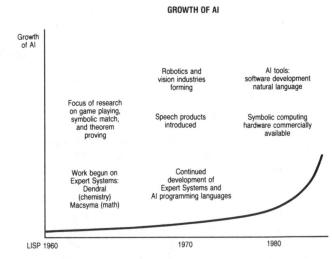

GROWTH OF AI

Fig. 1: AI research has been going on since the 1950s. Recent developments in software tools and computer hardware have caused a sudden increase in activity, and AI is beginning to move into the real world.

telligence, rather than machine-type processing. Machines, for example, are good at performing repetitive tasks, responding quickly to stimuli, and controlling great forces with precision. They can also store and retrieve vast amounts of numerical data. Humans, on the other hand, perceive patterns well, and can generalize, applying originality to conclusions and profiting from experience. Also, humans can improvise, exercise good judgment, selectively recall past experiences, and adapt to the unanticipated. AI tries to bridge the gap between humans and machines by giving the machines some aspects of human capabilities.

AI systems use computers in ways that are markedly different from those of conventional data processing. All computers deal with symbols, but traditional symbols usually represent mathematical equations and numbers. In contrast, AI computers handle symbols that can represent al-

— Symbolic representation of objects

Pump-23 is a centrifugal pump
Valve-01 is a gate valve
Valve-01 is connected to the input of pump-23

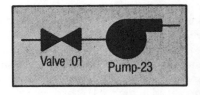

— Symbolic problem solving

- Input: What will happen if pump-23 is on and valve-01 is closed?
- Reasoning:
    + Valve-01 controls the input to pump-23
    + Pump-23 is a centrifugal pump
    + The input to a centrifugal pump must not be cut off
    + The input to pump-23 is not available
- Output: Pump-23 will seize if it is on and valve-01 is closed

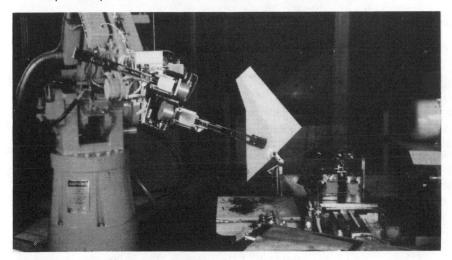

Fig. 2 (left, top): With rules and symbolic representations of facts, an AI system can reason that if Valve 01 is closed and Pump 23 is started, then Pump 23 will overheat and seize.

Fig. 3 (above): Voice input systems, such as the TI Speech Command System, are being used in many applications. This operator is examining a printed circuit board for defects and reporting results verbally.

Fig. 4 (left, bottom): Combining robots and vision systems is another area where much AI development is going on. Inspection robots are being developed that use an arm-mounted vision system to check parts and assemblies.

most anything—a concept, a person, a process, etc.

Attributes can be attached to objects and relationships established beforehand to create a very powerful symbolic approach to problem solving. For example, an AI system could symbolically represent a variety of pumps, valves, and their various interconnections (Fig. 2). It could then answer questions about the pumps and valves, such as what would happen if Pump 23 is turned on with Valve 01 closed. The symbolic reasoning process could deduce that the pump would eventually freeze up. In an actual installation, an alarm could be activated to warn of such a dangerous situation.

What impact will AI have in solving such real-world industrial problems? In general, AI will have the most effect in three areas:

- Man-machine interfaces;
- Robotics and vision systems;
- Tasks involving knowledge-based or "expert" systems.

Thus far, the man-machine interface has been improved through the use of color displays, high-resolution graphics, pointing devices, and touch screens. Speech synthesis and recognition (Fig. 3) greatly facilitate that communication and are being used more and more in industry. In the control room, for example, voice output systems are replacing bells and blinking lights. On the factory or warehouse floor, speech recognition systems are helping speed inventory control and quality inspection by eliminating the need to manually enter data.

Current AI research in voice I/O is aimed at expanding the capabilities of both speaker-independent and continuous speech systems—that is, systems that do not have to be trained to recognize a specific person's voice, and systems that can understand entire sentences.

Another man-machine interface area where AI is being used is *natural language.* Natural language processing allows users to interface to computers in English-like phrases instead of using rigid, meaningless computer commands. Using a common language, such as English, eliminates the need for learning many different commands for infrequently-used systems. One example is Texas Instruments' NaturalLink software for the TI Professional Computer. The software provides access to a number of popular database services, such as the Dow Jones News Retrieval.

Understanding English, however, requires an appreciation for the context in which it is used. Further, a system must be able to deal with the ambiguity that has crept into many English phrases. AI advances in these areas will allow greater vocabularies and more flexibility in accepting different phrases with the same meaning. Another goal of current AI research is a natural-language interface for control systems, such as those used for robot controllers.

Robots and vision systems

Robots and vision systems have captured the public's imagination through books and movies showing machines that mimic both a human's physical movements and visual perception. Unfortunately, the current state-of-the-art in robotics and vision is rather simplistic compared to their human counterparts. In fact, linking vision, perception, and consequent action is one of the most difficult tasks in AI research. This is partly because of limited understandings of how human vision and perception work, and how to orient machines with their external environment.

Nevertheless, robot development has gone ahead in such major application areas as welding, material handling, parts positioning, assembly, and spray painting. Such applications are most likely to be found in the automotive, electrical machinery, metalworking, electronics, aerospace, and plastics industries.

AI research in robotics is focusing on increasing a robot's flexibility, improving error recovery, enhancing its interface with people, and enabling it to learn as it operates. Development of sophisticated vision and tactile sensors is another key area for AI robotics research.

Like robotics, growth in machine vision has been spurred by industry's increasing focus on product quality. Until recently, most inspection tasks were performed manually. But there's a growing trend toward using vision for inspection operations, especially in automotive and electronics industries. At present, few vision systems are being used with robots. Instead, vision systems typically are used by themselves on factory lines, where they inspect, identify, and measure parts and assemblies.

Texas Instruments has made extensive use of both vision and robots. Using pattern recognition and image processing technologies developed at TI, inspection robots (which combine an arm and a vision system) can check parts and assemblies at a rate about three times that of human operators (Fig. 4).

Expert systems

The area of AI that has received the most publicity in recent years is that of knowledge-based Expert Systems. A problem-solving technique, an Expert System concentrates the collective processing power of public and private knowledge on specialized areas or domains. Thus, it can provide expert levels of performance in areas where human experts are either unavailable or not cost effective.

It should be emphasized, however, that Expert Systems are tools, not replacements for human labor. An expert system can, in many cases, free trained personnel to work in other areas. Also, key personnel sometimes retire, transfer, or quit; Expert Systems can smooth the transition and serve as a training tool for new personnel.

To construct an Expert System, a *knowledge engineer* first interviews a *process* or *domain expert* to gather the requisite knowledge (Fig. 5). Based on the initial interviews, the knowledge engineer selects knowledge-representation schemes and reasoning strategies. The data is then entered into the computer's knowledge-base compiler, which in turn translates it into the knowledge base.

The knowledge base in an Expert System is most often encoded in the form of *rules*. Rules contain two parts: an *antecedent*, which represents some pattern that must be matched; and a *consequent*, which specifies an action to be taken when the data matches that pattern. The antecedent typically contains several clauses linked by logical ANDs and ORs. The consequent contains one or more action phrases that specify what is to be done or what can be con-

cluded when the rule applies. A typical rule for an automobile diagnosis might be "IF having trouble starting engine, AND starter can crank engine, AND engine is receiving gasoline, THEN check for ignition problem."

Representing domain knowledge in the form of rules has several advantages. Humans understand and can relate to them as rules of thumb. Rules too are very modular and thus easy to change. Specifying rules is also non-procedural; that is, the system determines when to apply the rules. The rules also allow for an explanation of the system's conclusions, which is one of the hallmarks of an Expert System.

A block diagram of a typical Expert System includes an inference engine, a knowledge base, a workspace, a user interface, a reasoning explanation subsystem, and a knowledge acquisition subsystem (Fig. 6).

The inference engine is a computer program that processes rules stored in the knowledge base. Its control mechanism is kept separate from the knowledge base.

The workspace is an area of memory set aside for storing a description of the problem. The problem description can be constructed by the system from facts supplied by the user, or it can be inferred from the knowledge base during consultation.

The user interface is the communications link from the system to the user, and often entails, to some degree, natural-language processing.

The explanation subsystem keeps track of the reasoning steps in order to answer questions, such as how a particular conclusion was reached or why the system asked the user a particular question.

As noted above, the knowledge base contains rules and other data about the problem; the knowledge acquisition subsystem is the hardware/software used to enter that data. Typically, this is done via a terminal with menu-driven programs.

Typically, inference engines know how to process rules according to one or more *paradigms* or procedures. The two problem-solving methods most often used are *forward* and *backward inferencing*. In forward, or data-driven, inferencing, the system attempts to reason forward from the facts to a solution. For example, if X and Y are true, then so is Z. In backward, or goal-driven, inferencing, the system works backwards from a hypothetical solution to find evidence supporting that answer. An example of this is that if Z is assumed true, then it is necessary to find out if X and Y will support the conclusion. Expert Systems often combine these two methods.

Applications

Expert Systems have been used in a variety of successful applications. The most successful application has been in equipment diagnosis. The objectives of such systems are to reduce equipment downtime and maintenance costs by assisting technicians in the search for hardware problems. Although routine maintenance may not require the aid of an Expert System, finding failures in complex equipment often requires the guidance of an expert on the equipment's design and operation.

Expert Systems have been successfully used to trace problems in such equipment as locomotives, turbines, and programmable controllers. Similarly, diagnostic experts are helping investigate a range of application problems in

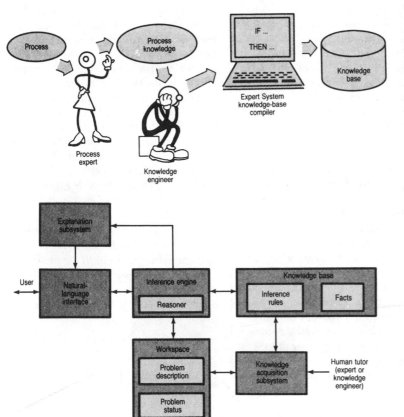

Fig. 5 (left, top): The first step in developing an Expert System requires building a knowledge base. This is done by asking question of a human expert, and entering the data as rules, antecendents, consequents, etc.

Fig. 6 (left, bottom): Block diagram of Expert System shows major components.

Fig. 7 (above): TI's Explorer system is an example of LISP-based computer hardware. Systems such as this will help get AI out of the research labs and into the field.

equipment as diverse as computers, jet engines, telephone cables, and cars.

Another application area is alarm analysis. These systems can help monitor a process in real time, interpret alarm conditions, and determine possible corrective actions. In fact, if the Expert System is tied into the process itself, corrective action can be taken automatically without human intervention. By decreasing reaction time, the losses incurred due to a mishandled alarm can be reduced or eliminated. (See the accompanying article for more on Expert Alarm Systems.)

On the production floor, Expert Systems can be used as operator aids, assisting workers in the operation of machinery. The systems provide a constant level of expertise at key operator stations, and a variety of status, production, and other information to assist plant management. New workers can benefit by the use of knowledge-based training tools. Natural-language interfaces, speech, and enhanced graphics allow the worker to proceed at his or her own learning speed.

Even before the plant is built, process planning Expert Systems can be used to lay out plant floor management and to simulate and diagnose processes. The cost of such systems can be easily offset by the reduction in planning time and elimination of scrap and less-than-nominal products. Once built, the plant can rely on knowledge-based scheduling systems to manage the variables involved in production process timing—set-up, inventory control, order entry, product mix, etc.

Besides aiding manufacturing and scheduling tasks, knowledge-based systems are being developed to improve the efficiency of the manufacturing process itself. These tools monitor and coordinate flexible manufacturing systems on a real-time basis, adapting to changes that occur on the plant floor.

Process optimization experts, guided by inference engines, will one day simulate processes and project trends. These systems will not only carefully monitor a manufacturing process and report on its status, but will react to undesired fluctuations in the real process. By interacting with sensors, they will be able to measure system performance, project its outcome, and tune the process for cost and quality advantage.

Tools of the trade

Because AI deals with very large and complex problems, the solutions require very powerful software and hardware tools. The development of those tools has let AI come out of the purely research-oriented laboratories and into the real world. For instance, several software packages are now available for the development of knowledge-based systems. The rule-based Personal Consultant from TI, for example, is a low-cost development tool and delivery vehicle that runs on the TI Professional Computer.

Two of the most significant tools to spin off from AI have been the LISP programming language, and the computers specifically built to use it. LISP (coined from the LISt Processing that is used extensively in AI) differs from other languages in that it is not algebraic or algorithmic in focus. Rather, it was developed specifically for symbolic processing. LISP can be interpreted on-line for interactive programming, or it can be compiled for increased performance. A major strength of the language is that it is extensible; that is, new functions can be defined by users and then added to the language. Moreover, problem-specific higher-level languages can be developed in LISP to run on top of the language itself.

But while the flexible data structures of LISP are extremely useful for representing knowledge, they are not handled efficiently by conventional computers. Because of this, special LISP processors have been developed to handle the symbolic data.

Special features of symbolic processing computers include a dedicated LISP processor, a large virtual address space, large amounts of physical memory, a high-resolution graphics display, and high-performance mass-storage devices. Although such features have previously been very expensive, new systems—such as the Texas Instruments Explorer comput-

er (Fig. 7)—have driven costs downward and given many more people access to such powerful tools.

The effect of low-cost LISP systems is two-fold. First, LISP machines will be used for general software development regardless of whether the end product will be written in LISP or not. The ability to rapidly prototype a software concept to see if it's worth further development should prove to be a valuable asset—one that will greatly enhance productivity. Second, as low-cost LISP computers become generally available, they can be used to deliver sophisticated knowledge-based systems to the plant floor.

Development is continuing on all aspects of symbolic processing hardware, even at the chip level. Texas Instruments, under government contract, is developing a custom LISP processor on a chip that will greatly increase the processing power of LISP-based machines.

The promise for AI is immense, but

a word of caution is in order too. AI has been around for a long time and has had some significant impact, but it is not by any means a mature technology. Yet, even with that warning, hopes are high. As the technology matures and costs decline, we can expect AI to diffuse widely. Over the long term, AI systems will increase our effectiveness, individually and collectively, by preserving know-how, distributing knowledge effectively, releasing workers from much of the boredom found in their jobs, and improving their performance. ∎

The authors, Dr. Richard A. Herrod and Barbara Papas, will be available to answer any questions you may have about this article. Dr. Herrod can be reached at (214) 995-0674 and Ms. Papas at (214) 995-4510.

CHAPTER 8

CASE STUDIES

Reprinted from *Aerospace America*, April 1985

With relevant modifications, new diesel-electric locomotive troubleshooting technology may yet help future aerospace systems stay on track

by **David I. Smith**
General Electric

CATS: precursor to aerospace expert systems?

Are expert systems *real*? Or is the whole thing made up and we are just hyping each other? Can aerospace companies —and people—really make a living in this market? Is the government—mostly through the Defense Advanced Research Projects Agency—throwing money away in a futile attempt to create an aerospace product for which no market exists?

Those are all tempting speculations but the fact is that expert systems *do* exist. Few are in aerospace, but they operate successfully where products must make a profit. Aerospace firms considering whether or not to take the plunge may now be challenged by the knowledge that such systems are here and *do* make economic sense.

One such expert system is GE's Computer-Aided Trouble-Shooting System (CATS) for diesel locomotives. It is operational, it is real, and it works. If an expert system can work for locomotives, why not for airplanes, or missiles, or spacecraft?

Upon leaving the maintenance shop a locomotive will run an average of 46 days (one-eighth of a year) before returning, be operated by about a hundred different engineers, travel over 20,000 mi., and use over 100,000 gal of fuel. A typical train has four units. Only the lead unit has an operating crew. Trailing units are controlled by electrical jumper lines running between units.

Operating personnel does not do any trouble-shooting, adjustment, or repair on the road. Instead a report is prepared for shop use during maintenance.

Reports are frequently vague. If the locomotive does not perform as expected, engineers will report deviations on a worksheet. One of the most common reports is "not loading properly." This is about the same as having a note taped to the refrigerator door saying, "Fix the car." You have no idea what the problem is—a faulty transmission, a plugged fuel filter, engine stall, a dead battery, or an empty tank.

Such reports confront the railroad maintenance shop with a difficult problem. Usually the shop checks over the locomotive and looks for some mechanical

David I. Smith is senior service engineer in General Electric's Locomotive Operation.

or electrical fault. Frequently none is found (perhaps because the unit had dirty lube oil filters that were changed on the service track on the way into the shop) and the locomotive is sent back into service. But if the problem complained about shows up during the check, the shop isolates the fault and fixes it. There is seldom much of a search for any *other* fault. This process has resulted in irretrievably lost time for the locomotive. It has also tied up shop space, and resulted in inefficient use of manpower. Until now.

GE's Erie, Pa., Transportation Division and its corporate R&D organization in Schenectady, N.Y., jointly worked on the CATS system for locomotives. This system stores a large number of "rules" representing the experience of a senior engineer. CATS reads these rules and asks questions via a video monitor screen, and receives answers through a simple keypad with keys labeled Yes, No, 0 through 9, Help, and Restart.

GE has fielded prototype CATS equipment for evaluation. The prototype, a rugged unit (packaged by Comark), contains a Digital Equipment PDP 11/73 computer that runs an RSX-11 operating system and an enhanced version of fig-Forth language, a 10Mb Winchester storage disk, a VT100 terminal, and a Selenar graphics board.

CATS will aid railroad shop personnel sift through symptoms for a pattern that may be called the "true" problem. CATS will in effect train maintenance crews in the proper analysis procedures that, in time, will enable them to find problems without computer assistance. In the case of experienced maintenance personnel, it will prompt them to ascertain that an obvious step has not been overlooked. A new shop employee can use CATS as a training tool for learning aspects of diesel locomotive technology ranging from underlying theory to the small details covered by Help aids.

If the computer does not have the information, it will display a question on a video monitor screen. The railroad maintenance man must answer the questions. The answer will be keyed in as Yes or No. In this manner CATS becomes increasingly informed of the locomotive's performance.

CATS's "rules" are conditional statements or questions expressed in railroad language. Each rule consists of a "situation" portion and an "action" portion (what to do if all items in a given situ-

Promising to transform even the most inexperienced locomotive mechanic into an instant "Mr. Fixit," GE's new computer-aided trouble-shooting system identifies causes of malfunction and demonstrates repair procedures.

ation are found to be true). When a rule is satisfied, an action specifies the next step required of the user and inserts this new fact into the computer's memory. The CATS interpreter works with facts and rules, monitoring facts in the data library and executing actions. It works forward trying to execute each rule and looks backward keeping track of various other objectives and verifying that it has enough supporting facts to meet each of them.

CATS's Help system offers extra assistance along the way. Whenever the Help key is pressed, a menu comes onto the screen, showing what form of help is available.

The mechanic may choose help from a videodisk player with a customized disk able to hold 54,000 separate color pictures

asked a seasoned troubleshooter if a certain indicating light came on at a particular point, and he told me that it did not, he would make sure that the bulb was not missing or burned out; he would test the bulb. If a computer program had been written to cover all phases of locomotive trouble-shooting, it would have to include a reminder to check the bulb each time an indicator light should be observed. With the expert system, the senior engineer only need create one new rule that says, "When an indicator light fails to come on, check the bulb." The expert system will thereafter issue the reminders without further software writing.

The CATS system is broken down into several major segments: engine mechanical, locomotive electrical, transmis-

---

EXCERPTS, TYPICAL DIALOG WITH CATS

REPORTED SYMPTOM IS: "NOT LOADING PROPERLY."

CATS: Is fuel oil pressure 40 psi or more, when measured with a test gage, with the engine running at idle?
Answer: No.
CATS: There is a fuel system fault. Does fuel-tank gage show enough fuel in tank to get good suction?
Answer: Yes.
CATS: Is fuel-transfer-pump motor running normally?
Answer: Yes.
CATS: Is unit equipped with a fuel filter drain valve?
Answer: Yes.
CATS: Is fuel filter drain valve closed?
Answer: No.
CATS: Close fuel filter drain valve. Confirm when done.
Answer: Done.
CATS: Now, is fuel pressure back to normal?
Answer: No.
CATS: Is fuel filter clean? Either check pressure drop, or verify that is was changed. If you don't know for sure, answer with "No."

Answer: Yes.
CATS: Is unit equipped with a fuel strainer?
Answer: Yes.
CATS: Is fuel strainer basket clean? Be careful of the O-ring in the lid.
Answer: No.
CATS: Clean or replace the fuel-strainer. Confirm when done.
Answer: Done.
CATS: Now, is fuel pressure back to normal?
Answer: No.
CATS: Is fuel-strainer gasket in good shape?
Answer: No.
CATS: Do not attempt to remove and clean O-ring. Usually O-ring swells slightly and cannot be put back successfully. Install new one. GE Part No. 115x1840. Replace fuel-strainer gasket. Confirm when done.
Answer: Done.
CATS: Now, is fuel pressure back to normal?
Answer: Yes.
CATS: Fuel system is normal. Fuel problem appears to be solved. Rechecking at full load must be done to be sure.

---

and two isolated sound tracks for presentation on a color monitor and speaker. Any frame can be brought up in less than 3 sec. This might be a single frame, a short training sequence with audio, or a single frame with associated audio. Help can also call up computer-aided design/manufacturing drawings on the CATS monitor. A drawing can also bring up text files and put them on the screen.

Help is not forced onto the user. Its services must be asked for and the Help process can be aborted at any time. One advantage of the CATS expert system can be illustrated by a simple example. If I

sion, engine auxiliary support, and auxiliary electrical systems. Users select a particular system from a menu, and can start at that point without going through the entire process.

With the aid of this new tool, GE hopes to give railroad maintenance people a means of doing proper repairs promptly, with reduced lost motion and more assurance that they have repaired the *last* problem on the locomotive, and not just the first one that they found. And the CATS experience may encourage the creation of expert systems for other complex machines.

By acceptance of this article, the Publisher and/or recipient acknowledges the U.S. Government's right to retain a nonexclusive, royalty-free license in and to any copyright covering this paper

Presented at the SME ULTRATECH Conference, September 1986

DORIS—A Case Study in Expert System Shell Development

by Keith L. Warn
Rockwell International Corporation

DORIS (diagnostic oriented Rockwell Intelligent System) was developed as an in-house expert system shell. This paper is a case study of the DORIS development and discusses: development decision making process, advantages/disadvantages vs. commercially available shells, design tradeoffs, and planned future enhancements. The planned future advancement discussed is an upgrade to Rockwell's manufacturing process in the area of production planning assistance.

DORIS (Diagnostic Oriented Rockwell Intelligent System) was developed as an in-house expert system shell. This permitted Rockwell the option to either select an in-house shell or a commercially available shell when proposing on contracts and when developing in-house projects, such as expert system managed automated manufacturing systems.

This paper discusses the following issues:
- DORIS development decision-making process
- DORIS advantages/disadvantages versus commercially available shells
- DORIS design tradeoffs
- DORIS planned future enhancements

One planned future enhancement discussed is the application of DORIS to upgrade Rockwell's manufacturing processes. One of these (tentatively being called MORRIS, for Manufacturing Oriented Rockwell Reliable, Intelligent System) is to upgrade Rockwell's production planning process at its Anaheim, California, facility. This process upgrade was selected because production planning constitutes a significant portion of manufacturing cost and because it involves judgment calls which cannot be easily automated using conventional software techniques.

INTRODUCTION

Expert systems, a branch of artificial intelligence, are finding increased applications in such fields as maintenance diagnostics and manufacturing. One reason for this increased application is that the expert system structure and support environment permits rapid prototyping. Also, expert system processing is more like human symbolic reasoning than that afforded by conventional processing. Situations where expert systems are preferable to conventional processing include those where the data is inexact, where some portions of the data are missing, where the logic results in a combinatorial explosion, and where the knowledge is continually being updated.

In practice, most large scale applications are actually hybrid systems which contain both expert systems and conventional processing. The expert system makes judgment calls based on rules and heuristics, and the conventional processor performs precision numerical processing (number crunching).

In a manufacturing environment, conventional processing would be used where straightforward, repetitive tasks are involved. Expert systems would be used for the larger applications, such as production planning assistance or production flow management assistance in which knowledge of the situation impacting the decision making process is not always exact, complete, or straightforward.

The key behind many expert system developments is the expert system shell. This shell usually consists of an inference engine, input/output functions, an explanation capability for decision traceability, a knowledge base upgrade feature, and supplier-unique support functions such as provisions to handle uncertainty in input data and conclusions.

The question that then arises in any expert system development is whether to develop and use one's own in-house expert system shell as the framework for the expert system or whether to purchase and use one of the many commercially available shells on the market. This paper describes the decision processes involved in Rockwell's development of its own shell and criteria for when this shell is to be used and when a commercially available shell is to be used instead.

BASIC PARTS OF AN EXPERT SYSTEM

Figure 1 shows the three basic parts inherent in any expert system. These parts are the current situation data, the knowledge base, and the inference engine.

The knowledge base contains the a priori domain knowledge by which the expert system decisions are made. These are usually in some form of IF... THEN... rules and reflect the heuristics (rules of thumb) that aid in making the decision.

The current situation data is that instant knowledge which is used to compare the knowledge base's rules against. In a manufacturing malfunction situation, this would include such information as the smell of electrical part smoke or a meter reading of 12 volts.

It is the inference engine that compares the current situation data, i.e., smoke and 12 volts, with the rules in the knowledge base, i.e., if smoke observed and 12 volts indicated on the digital multimeter, then resistor X_2 is bad with 0.82 certainty factor.

Because the knowledge base and the inference engine are separate, knowledge can easily be added or upgraded, and decision traceability can easily be changed along with changes in the knowledge base. Although a conventional system can certainly handle IF... THEN... expressions for simple problems, an expert system is the preferred approach when many rules are involved, when uncertainty needs to be reflected into the decision making process, and when the knowledge base must be continually updated as historical information is accumulated. An example of historical information accumulation would be experience gained over several years of a fielded piece of equipment that determined resistor X_5 to be the most likely cause of failure despite early design analyses to the contrary.

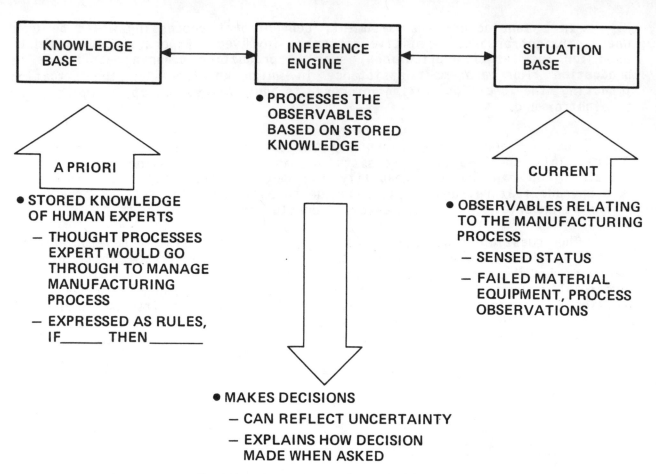

Figure 1. Key Elements of an Expert System.

REASONS FOR DEVELOPING OWN IN-HOUSE EXPERT SYSTEM SHELL

Before deciding to build our own in-house shell, a large number of commercially available expert system shells were evaluated. In early 1984, the moderate cost shells ($5K or less) evaluated were not very good. The expensive ($60K) shells were, on the other hand, determined to be quite good. After cost versus benefits trades, one expensive shell and several moderate cost shells were purchased. The expensive shell is now being used on a diagnostic oriented contract, and the moderate cost shells are no longer being used except for comparison and evaluation purposes.

Selection of commercial shells was found to be difficult because often the only information available was sales brochures and a briefing/demonstration by the vendor's sales force. It was difficult to make point by point comparisons between various shells, and new shells were coming on the market. This process was complicated by internal software procurement approval cycle delays, especially when the shells were in the expensive category.

Because suppliers of expert system shells normally do not release their source code, it was determined that there would be little chance for flexibility after committing to a commercial shell and that it would be difficult to build a commercial shell into a larger integrated system without making the larger system dependent on the strengths and weaknesses of the commercial shell. The problem here is that shell suppliers have a tendency to upgrade their software and to fail to support older versions after some period of time. Without access to the source code, it follows that a lot of money could be spent integrating a commercial shell into a product only to discover that the product must be upgraded to keep up with version updates of the shell just so that the vendor will support it. Because of the general nature of the commercial shells, many of these updates are not needed for the integrated system. Yet the product must be updated to be supported.

As a result of the above, the decision to develop the DORIS expert system shell with Rockwell IR&D funds was made in early 1984. The primary reasons behind this decision were as follows:

1. Suppliers of commercially available expert system shells normally do not release their source code. Hence, it cannot be modified without going back to the vendor. Also, one cannot get a hands-on understanding of how expert systems work without access to the source code.

2. As is standard practice for other software, there is usually a licensing fee for each additional computer on which the shell is hosted. With some of the better expert system shells on the market starting at $60K or more for the first copy, this can get expensive.

3. If one were to develop an expert system which would be applied to many computers, the multiple licensing fee could get out of hand. This is because an expert system developed on one shell is not normally executable on a shell developed by a different supplier. This problem of "shell addiction" is not likely to be resolved until expert system users (especially the Department of Defense) and shell developers agree upon some form of CRTIE (Common Run Time Inference Engine) standard whereby an expert system developed on one shell could be executed on a different shell which also conformed to the CRTIE standard[1].

DORIS ARCHITECTURE

Figure 2 shows the basic DORIS architecture. It has a priori knowledge bases, provisions for inputting the instant knowledge base, and an inference engine (inferencing systems).[2] It has both forward and backward chaining, where forward chaining starts out with the facts (instant knowledge) and comes up with a conclusion and where backward chaining hypothesizes a conclusion and determines whether the facts support such a conclusion.

The rule set executive interfaces with both the knowledge engineer and the end user. The knowledge engineer converts the domain knowledge (knowledge about the subject at hand) to rules in the knowledge base. This is done via the knowledge base maintenance system. The end user observes the decision and, if interested, can query DORIS to determine how the decision was made. The end user can supply instant knowledge, e.g., smoke observed, and can be asked questions by DORIS in a consultation mode.

Currently, Version 2.0 of DORIS uses production rules. The Version 3.0 upgrade, which is now in development, will include frames and variables to make it a more powerful shell.

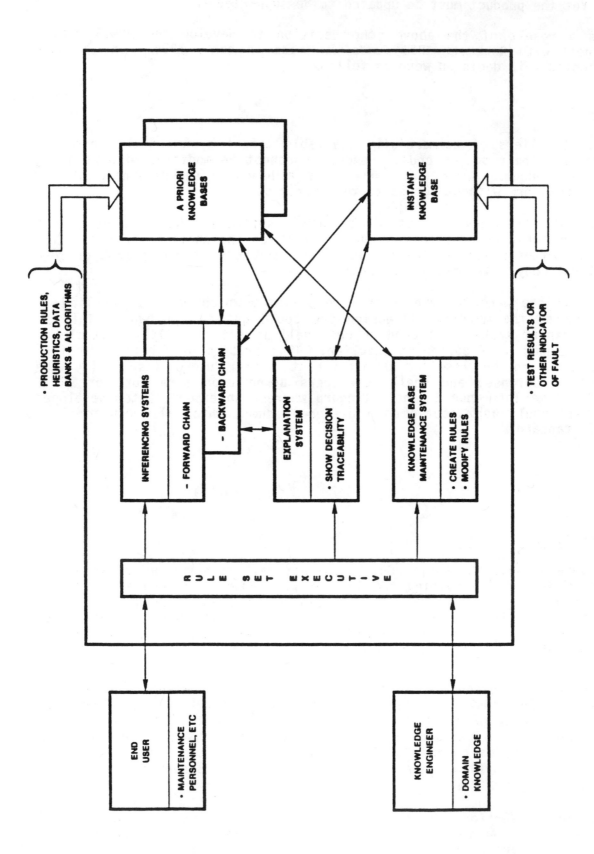

Figure 2. DORIS Architecture

DORIS EVOLUTION

Figure 3 summarizes the key events in the DORIS evolutionary process. When DORIS was first started, CommonLISP was not readily available. As a result, InterLISP was the first dialect used to code DORIS on a VAX 11/750.

When DORIS Version 1.0 was nearly complete, a sister division wanted a version which would operate for a navigation sequencing application on a personal computer. Since the DORIS code was too large for the PC, the backward chaining feature was deleted, and DORIS was converted to IQLISP, which was the only LISP dialect available on in-house PC's at that time.

DORIS was recoded into CommonLISP on a Symbolics 3640 as soon as CommonLISP was available on that computer, and some upgrades were made to result in a beta Version 2.0. When CommonLISP became available on the in-house VAX 11/785, it was transferred to that computer also. In theory, CommonLISP is supposed to be transportable. It was discovered, however, that it took about three hours of minor code changes to make it transportable from the Symbolics 3640 to the VAX 11/785. Conversion to CommonLISP early in the development of DORIS, it turns out, was a very good decision. This gave it the computer-to-computer transportability that some of the better and more expensive commercial shells still do not have today.

To increase the power of DORIS, the current activity is to upgrade the production rule based DORIS Version 2.0 into a frame-based version which includes variables. This option will permit inheritance of characteristics and will make it a more powerful fast prototyping tool.

Using this more powerful frame-based approach is not without some penalties, however. For example, all major frame-based expert system shell vendors were experiencing problems at that time with certainty factors because of problems caused by inheritance. The DORIS frame-based version may run into some of the same difficulties with uncertainty.

Another problem may be inferencing speed. For example, a comparison was made of the inferencing speed of DORIS with a $60K commercial shell being used on one of our contracts. DORIS took one second to perform a diagnostic evaluation on a benchmark case which the frame-based commercial shell took 26 seconds to perform. It originally was believed that the use of frames might have been the reason that the commercial shell was slower. Subsequently, it was observed in the literature that the speeds of structured production rule and frame based system are comparable.[3] An evaluation is now underway to determine if the difference in speed is due to different amounts of overhead.

Current plans are to compare the speed of the frame-based version of DORIS with the production rule version and to determine the causes of any speed differences observed. Since the approach taken clearly impacts inferencing speed, other organizations have deliberately developed their own expert system shells to optimize speed in applications like electronic warfare where decisions need to be made on the order of 10 to 100 milliseconds.[4]

Having access to an expert system's source code proved valuable for one reason never considered during the initial decision to build an in-house shell. In late 1985, the need existed to evaluate LISP vs. the DoD mandated Ada for embedded expert system applications.[5] Having access to an expert system's source code was valuable in the investigation. This study would not have been nearly as successful if only commercially available shell object code and no access to the source code was all that was available.

---

* Ada is a registered trademark of the U.S. Government (Ada Joint Program Office).

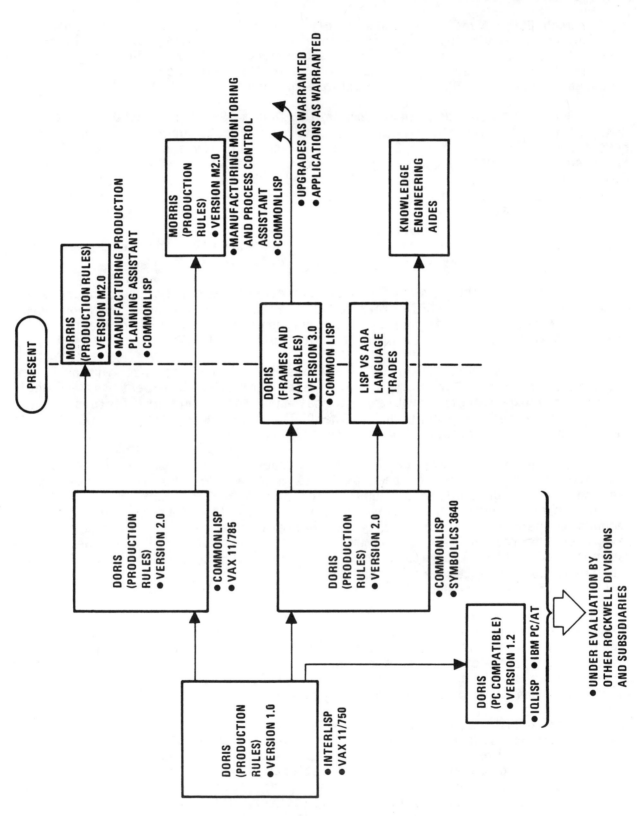

Figure 3. Key Events in Doris Evolution.

In early 1986, several manufacturing projects in which to apply artificial intelligence technology were considered. One of these included assisting the production manager in our El Paso, Texas, automated wire preparation and harness fabrication facility and another was for assisting in the production planning process.

The current automated wire system has a robot which handles some 35,000 different wires, some 100 different wire types, and 86 different end items. As envisioned, the expert system would have provided judgment call assistance to the human Process Control Manager.

A study of the current production planning process showed that the costs of production planning represented a reasonably large portion of manufacturing costs. For this reason, the production planning process was selected as the first application of artificial intelligence for manufacturing within Rockwell's Autonetics Strategic Systems Division. DORIS was selected as the shell for this system because it allowed access to the source code and because of its portability resulting from its being coded in CommonLISP. Because of potential changes to emphasize manufacturing, the manufacturing version of DORIS is being named MORRIS (Manufacturing Oriented Rockwell Reliable, Intelligent) so as to distinguish the two.

EXPERT SYSTEM-BASED PRODUCTION PLANNING ASSISTANT

Production planning has been a high cost driver on our products mainly because they are high tech products and because the production runs are not as large as would be found in mass production facilities like those for automobiles, tin cans, or milk cartons. The product to be manufactured is based on documents like specifications, process procedures, and drawings. Information from these documents are now converted by humans into production planning documents.

Of the domain knowledge required by human production planning personnel, some is of sufficiently low-level and medium-level skill that it could be handled by an expert system. Other domain knowledge is of such high level of skill or associated with a sufficiently high level of psycho-motor dexterity (like picking up a particular piece of paper on one desk and transporting it to another) that a human still needs to be in the loop. Since paper is a human-to-human or machine-to-human communications interface medium, the reduction of human participation in some steps offers the potential to reduce the amount of paperwork that it takes to manufacture a product.

Figure 4 shows how DORIS (now transformed into MORRIS) is planned to be implemented as a Production Planning Assistant. The a priori knowledge stored in the knowledge base will be that domain expertise relating to how to convert job orders into production plans. It will include product knowledge and knowledge about the material, equipment, and processes involved. The instant knowledge will include job order information and human inputs not available from the job order that is needed to establish the current situation.

The MORRIS inference engine will compare the instant knowledge (e.g., job order requirements) with the domain knowledge in the knowledge base. This domain knowledge will be similar to the thought processes a human would go through in converting job orders to production plans. The MORRIS output will be an interim production plan which would be checked and modified by the human as needed. The purpose of MORRIS' explanation system is to permit checking the logic behind the decisions made where the human might question or want to verify a particular portion of the interim Production Plan. After checked/ modified by the human (who could be a quality assurance person), the production plan would be released to the shop.

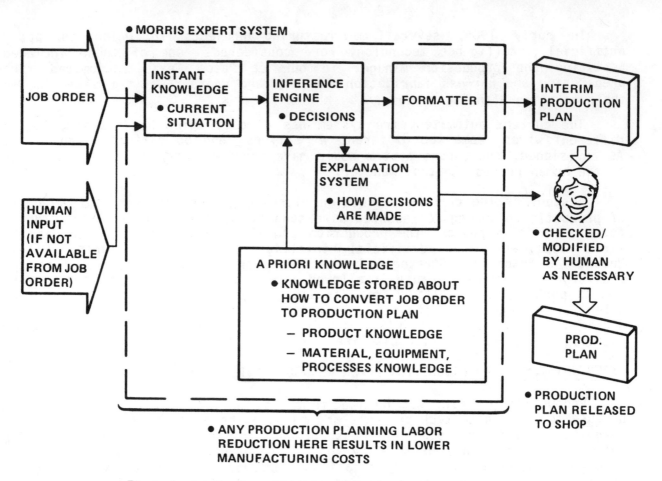

Figure 4. Application of Morris as a Production Planning Assistant.

LESSONS LEARNED

Rockwell benefited from the development of DORIS. The key lessons learned were:

1. CommonLISP is the best dialect of LISP to be used in expert system shells if transportability is of interest.

2. Development of an in-house expert system shell is a good way to get up to speed in the artificial intelligence field if a detailed understanding of the inferencing process is needed. This development, of course, requires company discretionary resources, and for a particular company or unit within a company, a trade needs to be made of knowledge benefit vs. cost.

3. A key advantage to development of an expert system shell is access to the source code; a disadvantage is that the more expensive commercially available shells usually have a better customer recognition value and image than in-house shells.

CONCLUSION

Development of an in-house shell has merit. The drawback is that the cost of development is more than the cost of purchasing a commercially available shell.

Lack of access to source code of commercially available shells is a major impediment to specialized applications, such as extremely high speed reasoning. The biggest problem, however, is that an expert system developed on one shell suffers from "shell addiction" and cannot normally be executed on a shell developed by a different supplier. The need exists, therefore for users and developers alike to get together and establish a CRTIE (Common Run Time Inference Engine) standard so that shell-to-shell portability can be achieved just like CommonLISP now permits computer-to-computer portability.

ACKNOWLEDGEMENTS

The author gratefully acknowledges the contributions made by Rockwell's Expert Systems and Software Engineering Unit in the development of DORIS, especially Kathleen Davis, Robin Webster, and Anne Stanley who were the principal early developers. Also, acknowledgement is made to Ron Meyer for his assistance in identifying how to apply DORIS to the production planning process.

REFERENCES

1. Warn, Keith L., "LISP Vs. Ada Implications in Diagnostics Oriented Expert Systems," Proceedings of IEEE AUTOTESTCON '86, September 9-11, 1986, San Antonio, Texas.

2. Davis, Kathleen D., "DORIS--Diagnostic Oriented Rockwell Intelligent System," Proceedings of IEEE 7th Annual Symposium - Integrated Diagnostics, December 5, 1985, Dayton, Ohio.

3. Niwa, K., Sasaki, K., and Ihara, H., "An Experimental Comparison of Knowledge Representation Schemes," The AI Magazine, Summer 1984.

4. Wright, M. L., "HEXSCON: A Hybrid Microcomputer-Based Expert System for Real-Time Control Applications," Proceedings of IEEE WESTEX '86, June 24-26, 1986, Anaheim, California.

5. DoD Directive 5000.31, "Computer Programming Language Policy," Department of Defense.

Presented at the SME EMTAS '84 Conference, October 1984

Intelligent Management System in Manufacturing

by Mark S. Fox

Carnegie-Mellon University

1. Laboratory Mission

The Robotics Institute at Carnegie-Mellon University was established in 1980 to undertake advanced research and development in seeing, thinking robots and intelligent machine systems, and to facilitate transfer of this technology to U.S. industry. The Intelligent Systems Laboratory within the Institute is concerned with extending and applying artificial intelligence techniques to problems in intelligent automation and robotics. The focus of our research to date has been in the area of intelligent systems to assist managerial, professional, engineering, production, and process control decision making.

The Lab employs 40 full and part time staff members and affiliated faculty under the direction of Mark Fox. Projects are underway to build both decision support and expert systems for managerial, professional, and production tasks. A perhaps unique aspect of the Lab is that while the work is divided into projects, the systems under development are designed to be interoperable with each other. Thus, projects will yield results that are not only useful in themselves, but which also can be combined into a larger, integrated package, which is called the Intelligent Management System. The following sections describe briefly our current projects.

2. The Intelligent Management System Project

A goal of the Intelligent Systems Laboratory of the Robotics Institute is to discover theories of knowledge acquisition, representation, and utilization that enable the construction an intelligent management system (IMS). An IMS must:

Sense: Automatically acquire state data. Sense the location of objects, state of machines and status of activities both on the plant floor and in supervisory departments.

Represent: Represent state, physical, functional, and historical information of objects and machines; tasks, goals, interests, etc. of people

Model: Model the complete organization at many levels of abstraction. For example, machines, people, materials, orders, departments, etc will be modelled in detail from both an attribute and a process view, including their interactions.

Manage: Analyze and manipulate the model to answer short and long term state and planning questions. The system is *passive* in that it responds to user initiated queries.

● Construct an active intelligent information system. Active systems continually

[1]Portions of this paper were authored by other members of the Intelligent Systems Laboratory, and is indicated in the footnotes.

This research is supported in part by Digital Equipment Corporation, Air Force Office of Scientific Research under contract number F49620-82-K0017, Westinghouse Electric Corporation, and the Aluminum Company of America.

monitor the organization and inform responsible personnel when important events occur. For example, when a machine break down occurs, not only is the foreman informed, but also maintenance, and the salesman who must inform the client that the order will be delayed.

Optimize: Analyze how the structure and the processing of the organization should be changed to further optimize some criteria such as cost, throughput, quality, etc.

Such a system, if it is to succeed in a business environment, must have the following characteristics:

Accessibility: Interfaces to computer systems are usually idiosyncratic and difficult to learn and use. Secondly, systems that change require that its users be continually re-educated. Our goal in the IMS is to enable all personnel to meaningfully communicate with the system. The interface will gracefully interact with the user and provide guidance and help in deciding what the user needs.

Accountability: A major hindrance in the acceptance of computer results in organizations, is the inability of users to question how the results were generated. Our goal is to construct an explanation system which will allow the IMS to explain its actions at various levels of detail to the user.

Adaptability: Currently, software systems are tightly coupled, requiring extensive programming whenever changes are required. This has resulted in the growth of the need for programmers in creating and maintaining organization computer systems. The goal of our research is to construct a theory of system design which will allow users to modify the model, analysis and processing functions, etc. in the system without the aid of programmers. The end user introduces changes via dialogue.

Reactability: The system, via its sensors and data monitoring is able to detect changes in the organization. Detrimental changes are corrected if possible. Interested personnel are informed of the change and only received the information they require.

Reliability: The system will not fail if one of its parts fails. No resource is critical.

3. Organization Modeling
[2]The management and analysis of an organization requires a richness and variety of information not commonly found in the databases of management information systems. For example, a simulation system requires knowledge of existing processes including process times, resource requirements, and its structural (routing) relation to other processes. It must also know when routings for products are static, or are determined by a decision process such as a scheduler. In the latter case, it must know when and where to integrate the scheduler into the simulation. If the IMS is to generate the sequence of events to produce a new product, it must have knowledge of processes (e.g., machines) which includes the type of processing it can do, its operating constraints, the resources it consumes, and its operating tolerances. If data is to be changed in an interactive, possibly natural language mode, the IMS must have knowledge of generic processes such as machines, tasks, and departments if it is to understand the interaction. It must also know what information is important and how it relates to other information in order to detect missing information and inconsistencies. Hence, the organizational model must be able to represent object and process descriptions (structural and behavioral), and functional, communication and authority interactions and dependencies. It must represent individual machines, tools, materials, and people, and also more abstract concepts of departments, tasks, and goals.

---

[2]This research is performed by Mark Fox and Stephen Smith.

Current organizational models are found typically in databases fragmented across one or more computer systems. How information in the database is interpreted is defined by the program and not by agreed upon conventions of field and relation names (though work in relation schemata is proceeding). By taking an AI knowledge representation approach to organization modelling, the variety of information described above can be represented. The model is accessible by all subsystems while the semantics of the model is jointly understood. Secondly, an AI approach to organization modelling provides the information required by all the management and analysis functions.

While many of the information enumerated above can be represented using current AI knowledge representation techniques, there is still much that requires craftsmanship and is poorly understood. More work is required to standardize the representation of causal relations, data changes over time, and idiosyncratic inheritance relations.

To date, our research has focused on the use of the SRL knowledge representation system as the basis for organization modeling. SRL has been extended to include conceptual primitives such as:

- actions, states, and objects,
- constraints and their relaxations,
- time,
- causality and dependencies, and
- belief.

With these primitives, detailed models of over 5 plants have been constructed.

4. ISIS Production Management and Control System
[3]

4.1. Introduction

The ISIS project began in the summer of 1980 in conjunction with the Westinghouse Corporation Turbine Component Plant in Winston-Salem N.C. The goal of ISIS is to investigate new, artificial intelligence-based approaches to solving problems in the management and control of production in a job-shop environments. The results of this investigation being an operational prototype.

At present three versions of ISIS have been constructed: ISIS-I (December 1980), ISIS-II (December 1981), and ISIS-II.4 (December 1982). ISIS has been designed and implemented so that its functions are independent of the particular plant. The remainder of this summary describes the major capabilities of the latest version.

4.2. Organization Modeling

The level of intelligent processing behavior a system may exhibit is limited by the knowledge it has of its task and its environment. To enable ISIS to perform "intelligent" management and production control, an artificial intelligence approach is used to model the production environment. The SRL knowledge representation system (Fox, 1979; Wright & Fox, 1982) is used to model all relevant information. Conceptual primitives have been defined for the modeling of organizations. They include:

- States (of the organization),
- Object Descriptions (e.g., parts, attributes),

---

[3]This research is peformed by Ranjan Chak, Mark Fox, Stephen Smith, Gary Strohm, and Douglas Zimmerman.

- Goals (e.g., shipping orders),
- Time,
- Causality, and
- Possession.

4.3. SRL

SRL provides ISIS with the capability of modeling a plant at all levels of detail; from physical machine descriptions, to process descriptions, to organizational structures and relations. SRL subsumes functions normally provided by database systems. ISIS uses SRL to describe all products, resource including machines, tools, personnel, etc., operations, departments, plant layout, and other information necessary to support its functionality. The following describes some of SRL's uses in ISIS:

- *Order definition.* Any type of order, e.g., live, forecast, customer, manufacturing, etc., can be created, and updated interactively. New types of orders can be created as needed.

- *Lot definition.* Orders may be grouped into lots, which may be run as a unit through the plant.

- *Resource definition.* Resources such as machines, tools, fixtures, materials, and personnel can be defined and used by an extensible set of functions. Resource definitions include substitutability in operations, and their current operation assignments in the plant.

- *Lineup (operation) definition.* How a product may be produced may be described as an operations graph. The operations graph describes all alternative operations, processing information, and resource requirements. Operations can be described hierarchically, enabling the description of operations at varying levels of detail.

- *Work area definition.* Cost centers, work areas, and any other plant floor organizations may be defined and resources, sub work areas, etc. can be assigned to them. Possible uses besides scheduling include accounting, personnel, and other functions.

- *Department definition.* Departments, personnel, and any other organization structures may be defined, and linked with other parts of the model.

- *Reservation definition.* Any resource may be reserved for an activity (operation). ISIS provides full reservation creation and alteration both interactively and automatically (see scheduling).

- *Plant organization.* The plant may be described hierarchical both from an organization structure perspective, and a physical layout perspective. This is used to support functions such as color graphic displays of the plant layout.

4.4. Model Perusal

ISIS provides full interactive model perusal and editing from multiple users in parallel. The user may alter the model of the plant from any terminal. Perusal is providing in a number of forms including menus, a simple subset of english, and graphic displays.

4.5. Interactive Scheduling

The interactive sub-system of ISIS provides the user with the "hands on" capability of creating and altering schedules interactively. The system monitors the users actions and signals when constraints are broken. The following are a few of its features:

- *Lotting.* ISIS provides the user with an interactive lotting facility for searching and examining orders, and grouping them into lots.

- *Resource Scheduling.* Resources may be scheduled by reserving them for use in a particular operation at user specified time. Such resources are noted both with the resource and the reserver (lot).

- *Hierarchical scheduling.* The user may be construct schedules at different levels of abstraction, and ISIS will automatically fill in the other levels. For example, the scheduler may schedule only the critical facilities, and ISIS will complete the schedule at the detailed lineup level.

- *Overlap flagging.* If user defined reservations for resources result in conflicting assignments, ISIS will inform the scheduler, and may automatically shift other reservations.

- *Constraint checking.* Whenever a reservation for a resource is made, ISIS will check all relevant constraints, and inform the user of their satisfaction. For example, if the reserved machine cannot be used for that sized product, the user will be informed at the time the reservation is made. ISIS allows the specification of almost any constraint the user may wish to specify (see automatic scheduling).

4.6. Automatic Scheduling

Analysis of the Winston-Salem plant, and other job-shops, has shown that the driving force behind scheduling is the determination and satisfaction of constraints from *all* parts of the plant. Current approaches to scheduling fail in their inability to consider *all* the constraints found in a plant, hence their results are mere suggestions which are continually changed on the factory floor. Hence ISIS was designed to perform constraint-directed scheduling of job-shops. ISIS provides the capability of defining and using almost any constraint in the construction of schedules. It is able to select which constraints to satisfy, on an order by order basis, and if they are not all satisfiable, ISIS may relax these constraints. Types of constraints include:

Operation alternatives	Operation Preferences
Machine alternatives	Machine Preferences
Machine physical constraints	Set-up times
Queue ordering preferences	Queue stability
Start date	Due date
Work-in-process	Local queue time
Resource requirements	Resource substitutions
Personnel requirement	Resource reservations
Shifts	Down time
Productivity	Cost
Productivity goals	Quality
Order priority	Lot priority

The following are some of the capabilities of automatic scheduling:

- *Constraint Representation.* Much of the work in ISIS has centered around a general approach to the representation of constraints. The results of this research allows ISIS to represent and use almost any constraint the user desires. Information representable in a constraint includes: duration of applicability over time and during operations, obligation by the system to use the constraint, context of applicability, importance of the constraint, relaxations of the constraint if it cannot be satisfied, interactions amongst constraints, and the utility of their satisfaction.

- **Multi-level Scheduling.** ISIS allows scheduling to be performed at differing levels of detail and perspectives. It currently performs a bottleneck analysis whose output is a set of constraints to the detailed scheduling level.

- **Bottleneck Analysis.** The capacity analysis level of scheduling determines the availability of machines to produce an order. It outputs a set of constraints on when each operation for an order should be performed so that it can avoid unnecessary waiting, and tardiness.

- **Detailed Scheduling.** The detailed level of scheduling provides complete scheduling of all resources required to produce an order. It takes into account all relevant constraints both model and user defined. As it constructs a schedule, it tests to see how well the schedule satisfies the known constraints. If important constraints cannot be satisfied, they will be relaxed. This level searches for a constraint satisfying schedule.

4.7. Reactive Plant Monitoring

ISIS provides the user with interfaces to update the status of all orders and resources. If the new status does not coincide with its schedule, then ISIS will reschedule only the affected resources. For example, if a machine breaks down, then all affected orders are rescheduled. Hence, ISIS can be used to provide realtime reactive scheduling of plants. It can also be extended to connect to an online data gathering system on the plant floor, providing realtime updating of plant status.

4.8. Generative Process Planning

One of ISIS's features is that its constraint representation allows it to perform a subset of generative process planning. Currently, ISIS has knowledge of a product's basic physical characteristics, and can choose machines based on them. These constraints can be extended to include geometric information.

4.9. Resource Planning

Since ISIS schedules all specified resources, the user is informed of the resource requirements needed to satisfy the production. Constraints on the utilization of resources (e.g., machines, tools, personnel, etc.) may be specified and used to guide detailed scheduling. This enables departments, like advance planning, to specify resource utilization constraints directly to ISIS.

4.10. Accessibility: Flexible Interfaces

The ISIS interface is menu/window based. The user is presented with multiple windows of information on the screen, plus a menu of commands to choose from. The menu system provides a network of displays ranging from order entry/update to interactive lotting to report generation. All reports may be printed directly to the screen or to an attached printer. A simple natural language interface is also provided to the user for perusing the plant model. A device independent color graphics display system is also available to ISIS. It is currently used to view a blueprint-like display of the plant, and to zoom in on work areas and machines. It can be used to display the status of the plant during operation.

ISIS has been designed to be a multi-user system. Its model can be shared amongst multiple programs. Hence, the model may be perused and altered from multiple locations in the plant, allowing departments to get the information they need to perform their tasks, and to provide information directly to ISIS. ISIS is being extended to determine who the user is and restrict access and alteration capabilities depending on their function.

4.11. Simulation

ISIS is part of the Intelligent Management Project, hence can use other functions available in the project. For example, KBS, a knowledge based simulator, can interpret the organization model directly to perform simulations. Since ISIS already has a model of the plant, KBS can use that model to perform simulations. All the user has to do is modify the model to reflect environment they wish to simulate.

This research is sponsored by the Air Force Office of Scientific Research, the Westinghouse Electric Corporation, and the Robotics Institute.

5. I-NET: A Knowledge Based Simulation Model of a Corporate Distribution System

[4] The I-NET project is an application of Knowledge Based Simulation (KBS) techniques to the domain of corporate distribution management. Corporate distribution management provides a rich environment for studying new techniques developed in KBS. Consider a typical manufacturing organization which manufactures a number of products and whose components are manufactured in a number of widely separated locations. These components are warehoused and merged at different locations and distributed to reseller locations. In such a system there are numerous decisions that have to be made about the transportation, warehousing, manufacturing and order administration policies. The purpose of I-NET is to provide a simulation model which can be understood, modified and used by managers directly without the assistance of a programmer. These facilities should provide the manager with an indepth understanding of the distribution network and aid in decision making.

Currently the I-NET system consists of:

- A knowledge based network editor to create the corporate distribution network

- A demand editor to model demand distribution

- A map perusal facility to graphically display the corporate distribution network

- A report generator to facilitate comparison of the performance of various scenarios

Two newly created facilities of KBS: model management and graphic display of model dynamics are being applied to the I-NET project in order to study their utility and determine future directions for research.

6. ROME: Knowledge-Based Support Systems for Long Range Planning

[5] *The problem: supporting long-range financial planning* Many business decisions are based on information produced by computerized financial planning models. While the models themselves may be quite sophisticated, their computer implementations generally do little more than calculate and display the results. Not much attention is given to screening the input data for anomalies, verifying that the data satisfy the assumptions of the model, or checking to make sure that the outputs seem reasonable for the situation at hand. Nor are there facilities for explaining what the outputs represent, showing their derivation, or justifying the results to users who are not familiar with what a particular program does. Traditionally, these tasks have been left to human analysts who could intelligently apply a programmed model to answer managerial questions.

---

[4] This research is performed by Mark Fox, Nizwer Husain, Malcolm McRoberts, and Ramana Reddy.

[5] This research is performed by Dave Adam, Brad Allen, Don Kosy, Peter Spirtes and Ben Wise.

ROME: a Reason-Oriented Modeling Environment The ROME project, being sponsored by the Digital Equipment Corporation, is an effort to develop a knowledge-based system which could itself perform many of the above tasks and hence more effectively support decision-making in the area of long-range financial planning. Our approach is based on the idea that current programs are limited by a lack of knowledge, i.e., they simply don't know what the variables in the models they manipulate mean. For example, they don't have knowledge of how the variables are defined in terms of real-world entities and so they can't explain what the variables stand for. They don't themselves keep track of the relationships used to derive the variables and so they can't explain how they got their values. They have no knowledge of "normal" versus "abnormal" circumstances and so cannot detect peculiar values, whether they be for input, intermediate, or output variables. Finally, they have no sense of the consequences implied by the variables and hence cannot tell "good" values from "bad" ones with respect to the goals of the organization.

In contrast, our overall goal for ROME is to make the meaning of the variables available to and usable by the system itself. Therefore, we have developed an expressive representation for financial models using the SRL1.5 knowledge representation language. This representation allows ROME to keep track of the logical support for model variables, such as their external source, method of calculation, and assumptions that must hold for the variable's values to be valid. Tracing back through the dependencies associated with a variable's computation can be used to explain why a value should be believed. Similarly, ROME can challenge the values of a particular variable by comparing them against relevant expectations, organizational goals, and independently derived values. Prototype implementation of two ROME subsystems, called ROMULUS and REMUS, has recently been completed.

ROMULUS: Reason-Oriented Modeling Using a Language Understanding System. ROMULUS is the user interface for the ROME system. Instead of the rigid and stylized input language used with most computerized support systems, ROMULUS has been designed to accept natural language queries about the model expressed in English sentence form. The query types currently understood are those which relate to definitions and calculations such as "What is the definition of production spending?" and "How was line 46 calculated?" ROMULUS also supports the interactive construction and editing of financial models in natural language, by allowing the addition of new variables, formulas and constraints on variables. Examples of acceptable user assertions are "Define year end people to be direct labor + indirect labor", and "Expect direct labor to go up." A major goal for ROMULUS has been to make the system as cooperative as possible, by including ways to recover from user mistakes (e.g. by spelling correction) and to tolerate user ignorance (e.g. by accepting synonyms and variations in syntactic form).

REMUS: Reviewing and Analyzing a Model's Underlying Structure. REMUS is the financial model reviewing expert for the ROME system. Given a financial model and a set of constraints entered by users which represent plan reviewer expectations and corporate goals, REMUS scans the model to detect constraint violations, which are then reported to the user. When a constraint violation is detected, REMUS attempts to determine the underlying circumstances that account for it by examining the formulas, input and intermediate variables that involved in it's computation. By this process, REMUS can localize the source of a constraint violation to the input variable(s) which seem to be responsible.

Current status An integrated version of the ROME system, called ROME1.0, was delivered to Digital in October of 1983. It is currently undergoing testing and development at the Digital facility in Marlboro, Massachusetts. We are presently involved in extending the capabilities of the ROME system in three areas: the causal diagnosis of constraint violations, the dependency-based revision of financial models in the face of inconsistency or change, and the support of user exploration of hypothetical plans. This work will involve major additions to not only the REMUS and ROMULUS subsystems, but to the SRL1.5 knowledge representation language itself.

7. PTRANS: Manufacturing and Distribution Management

[6] *Application Description:* In many manufacturing environments, events that disrupt production plans occur fairly frequently (e.g. customer change orders, material or capacity shortfalls, delays in new product introduction). PTRANS is a system that generates plans for the manufacture and distribution of computer systems. When an event occurs that makes one of these plans unimplementable, it revises the plan and any other plans affected by that revision.

Basic Research: The basic research issues that our work on PTRANS explores are:

1. understanding how to propagate the implications of an unexpected event across a set of plans in a way that minimizes the disruptive effect of that event, and

2. understanding how to justify plans in a way that can be easily understood by users of the system.

Status: PTRANS actually consists of two closely couples systems, IMACS (which generates plans for the assembly and test of computer systems) and ILOG (which generates plans for the distribution of these systems to the customer). Imacs, has begun to be tested at a Digital facility. A prototype version of ILOG will be delivered to Digital in June, 1984.

8. CALLISTO -- A System for Managing Large Design Projects

[7] Innovation is playing an increasing role in the continued vitality of industry. New products and innovations in existing products are occurring at an increasing rate. As a result, product lives are ever decreasing. In order to maintain market share, companies are being forced to reduce product development time in order to enter the market. By entering the market as early as possible, the ever decreasing product life will be extended.

A major portion of the development cycle is consumed in the performance and management of activities. For example, in high technology industries such as the computer industry, thousands of activities are required to be performed in the design and prototype build of a new product. Poor performance or poor management of an activity can result in critical delays. If product development time is to be reduced, then better management and technical support should be provided to each of the activities.

The Callisto project examines the extension and application of artificial intelligence techniques to the domain of large project management. Managing large projects entails many tasks, including:

Plan Generation and Scheduling: selecting activities and assigning resources to accomplish some task.
Monitoring and Control: monitoring the status of parallel activities in order to ascertain both plan and schedule changes required to meet project goals.
Product Management: maintaining a current description of the product (which is usually the outcome of a project), and determining the effects of changes to its definition (e.g., engineering change orders).
Resource Management: acquisition, storage, and assignment of the many resources required to support a project.

A close observation of project tasks shows that errors and inefficiencies increase as the size of the project grows. The successful performance of project tasks are hindered by:

---

[6]This research is performed by Paul Haley, and John McDermott.

[7]This research is performed by Mark Fox, Michael Greenberg, Joe Mattis, Drew Mendler, Tom Morton, Mike Rychener, and Arvind Sathi.

Complexity: due to the number and degree of interactions among activities (e.g., resources, decisions, etc.).

Uncertainty: of direction due to the unknown state of other activities and the environment.

Change: in activities to be performed and products to be produced, requiring project flexibility and adaptability.

While CPM and PERT techniques provide critical path and scheduling capabilities, the bulk of the tasks are performed manually.

Callisto provides decision support and decision making facilities in each of the above tasks. The ability to extend the capabilities found in classical approaches is due to Callisto's project model. Starting with the the SRL knowledge representation language, a set of conceptual primitives including time, causality, object descriptions, and possession are used to define the concept of activities and product. The language is further extended by the inclusion of a constraint language, representing the constraints amongst activities. The modeling language provides Callisto with the ability to model both products and activities in enough detail that inferential processing may be performed.

Callisto's decision support and decision making capabilities include:

- interactive change order management for products,
- multi-level scheduling of activities,
- rule-based analysis and maintenance of activities, and
- automatic generation of graphic displays of project models.

These functions are constructed from a combination of three problem-solving architectures:

Event/Agenda Based: Callisto can interpret a user's process, represented as a network of activities and states, by setting up and maintaining an agenda of goals and monitored events. This processing facility is used by the scheduling algorithms.

Rule-Based Programming: PSRL, a production system language built on top of SRL, is used to implement managerial heuristics for project management. PSRL can monitor and act on arbitrary SRL conditions.

Logic Programming: HSRL, a Horne clause theorem prover, is used as a question answering mechanism. HSRL represents assertions and theorems within SRL.

Current research focuses on a distributed Callisto system. Each member of a project has a "mini-callisto" to aid in managing their task. Each mini-callisto is able to communicate with other mini-callisto's to collectively manage the entire project.

A version of Callisto is currently being tested at Dec. Callisto is supported by Digital Equipment Corporation.

9. Factory Monitoring System

[8] The problem which the Factory Monitoring System addresses is the need to monitor and control a factory, giving supervisors dynamic access to the factory floor. The system must not only provide control and monitoring but allow analysis. The focus of this research is to implement a distributed process system providing supervisors with user interface processes, UIP's, from which they can access data about the factory floor.

A multiprocess system uses the basic technique of a model independent simulation augmented by

---

[8] This research is performed by Mark Fox, and Drew Mendler.

simulating each machine as a separate process on the computer; specifically the subject of this simulation is the HAP3 lamp prodution group of Westinghouse's Fairmont, West Virginia fluorescent lamp plant. Since each machine is a separate process a general protocol was designed for communication from a machine process to other machine processes and to the user interface process. The user interface, being a separate process, can devote all it's resources to interacting gracefully to request and information handling. The facilitation of information track and data collection throughout the system was done by passing, from machine process to machine process, a knowledge representation based description of the particular lamp.

General Communication Protocol

The use of a general communication protocol between a machine and the user interface process enables the information received to be decoupled from the data source. Requests sent to a machine are also done using the communication protocol. The general communication protocol along with the knowledge representation language's ability to describe each state in the production process facilitated the process control of lamp through the system.

Natural Language Interface

The ability of the user to interact by full or fragmented english sentences provides a graceful interface for process monitoring and control of the system. The use of a natural language interface adapts to the needs of different users by giving them a fairly diverse subset of english sentences.

Information Tracking and Data Collection

In the current system, tracking of a lamp is performed by creating an instance relationship between the specific lamp, a LIS, and the expected characteristics of a lamp on that machine, the LES. Information about the time dependent information for this unique lamp are gathered together in the LIS. Defect tracking is also handled in a similar way. Whenever a lamp is found defective, the machine's LES count of defects is incremented which triggers a signal to be sent to the user interface process. This method provides a way of notifying the user of important events in the system via the communication protocol.

10. WASTE: Process Diagnosis and Simulation

[9] The WASTE project is an investigation into the use of causal models for the purpose of process diagnosis and design. The domain is the treatment of chemical wastewater. During operation of the treatment system pipes and valves clog, pumps break and sensors fail. While shallow fault models of the system might tie symptoms of these failures to the fault, they cannot operate robustly. When the problems are unusual or complex the shallow models are of little use.

We are also interested in sharing models between multiple applications. Models are difficult and expensive to both build and maintain. If a single model can be constructed which can be used for multiple functions, efficiencies have been achieved. Thus, a simulation model constructed to aid in design of a process which can be used later to also diagnosis process faults is of great value. A single model can live on with the system.

Hence, complex, quantitative simulation models of the kind used for process design and analysis are taken as a starting point for the diagnosis system. These are represented as SRL models and augmented by representations of the physical structure of the process' environment. Diagnosis is then performed based on these deep models. The complexity of these models, however, requires that a variety of tools be brought into

[9]This research is performed by Mark Wright.

play to make diagnosis possible. First, the simulation models typically consist of real valued functions. There are literally an infinite number of different configurations of these model's parameters that could be explored to find one which explains the process' faults. To abstract away some of the information, a qualitative interpretation of the simulation models is taken. The actual real-valued functions are transformed into equations which reflect only how the direction of change of one parameter effects the direction of change of other parameters.

Unfortunately, this still leaves a great number of diagnosis possibilities. Hence, further abstraction of the model is performed. The system is represented at multiple levels of abstraction to allow diagnoses to take place first on the simpler abstract models. As diagnoses are hypothesised, they are further tested at more detailed abstraction levels.

As a final tool, belief knowledge is used to help focus the search for correct diagnoses. Components that have a history of failure are noted as such and faults involving these components are be explored more quickly than those which are usually functioning. As the more common faults are exhausted, less and less likely faults are explored.

11. The ALADIN Alloy Design System
[10] Alloy design is a metallurgical problem in which a selection of basic elements are *combined* and *fabricated* resulting in an alloy that displays a set of desired characteristics (e.g., fracture toughness, stress corrosion cracking). The quest for a new alloy is usually driven by new product requirements. Once the metallurgical expert receives a set of requirements for a new alluminum alloy, he/she begins a search in the literature for an existing alloy that satisfies them. If such an alloy is not known, the expert may draw upon experiential, heuristic, and theory-based knowledge in order to suggest a set of new alloys of which one may result in a successful combination.

The quest for a "successful" alloy may require many hypothesis/experiment cycles, spanning several years. Due to the amount of information published each year, it is impossible for an expert to keep track of each new development in other labs, and even in their own lab. Hence, some hypotheses might have been rejected if the expert had been aware of previous experiments performed either in the same lab, or in other corporations or universities. If it were possible to provide more information during the alloy design phase, the number of false hypotheses would be reduced, allowing the search to converge more quickly and at a lower cost. In addition, not all alloy design experts are created equal. Some are more "expert" than others, and their expertise covers different areas of knowledge. Capturing some of the knowledge used by a variety of experts to design alloys, in an accessible form, would extend the design powers of others.

The goal of the Aladin project is to perform research resulting in the design and construction of a prototype, AI-based, decision support system for aluminum alloy design. In particular to develop a knowledge bank of alloy knowledge, and to construct a problem-solving capability that utilizes the knowledge bank to suggest and/or verify alloy designs.

Knowledge Bank. The system will act as a database of all alloy-related knowledge necessary to design new alloys. It will have interfaces which will enable the perusal of information, and the addition of new information.

Design Aide. The system will aid in the suggestion and verification of alloy design. Based on heuristic and analytic knowledge found in experts and the literature, a problem-solving capability will be constructed which, when provided with a set of alloy requirements, will either suggest a set of hypotheses with a certainty

---

[10]This research is performed by Mark Fox, Ingemar Hulthage, and Mike Rychener.

rating, or assign a certainty rating to a hypothesis provided by the user.

This research is supported by the Alcoa Corp.

12. Laboratory Publications

Allen, B.P., and Wright, J.M. "HSRL: An Approach to Hybrid Knowledge Representation." Technical Report, Robotics Institute, Carnegie-Mellon University, Pittsburgh, Pennsylvania, to appear.

Allen, B.P., and Wright, J.M., (1983),"Integrating Logic Programs and Schemata." *Proceedings of the 8th International Joint Conference on Artificial Intelligence*, Karlsruhe, West Germany, 1983.

Fox M.S., (1979), "On Inheritance in Knowledge Representation", *Proceedings of the Sixth International Joint Conference on Artificial Intelligence*, pp. 282-284, Tokyo Japan.

Fox M.S., (1981), "An Organizational View of Distributed Systems", *IEEE Transactions on Systems, Man, and Cybernetics*, Vol. SMC-11, No. 1, January 1981, pp. 70-80.

Fox M.S., (1981), "The Intelligent Management System: An Overview", Technical Report CMU-RI-TR-81-4, Robotics Institute, Carnegie-Mellon University, Pittsburgh, PA, July 1981.

Fox M.S., (1981), "Reasoning With Incomplete Knowledge in a Resource-Limited Environment: Integrating Reasoning and Knowledge Acquisition", *Proceedings of the Seventh International Joint Conference on Artificial Intelligence*, Vancouver BC, Canada, Aug. 1981.

Fox M.S., (1983), "The Intelligent Management System: An Overview", In *Processes and Tools for Decision Support*, H.G. Sol.(Ed.), North-Holland Pub. Co.

Fox M.S., (1983), "Constraint-Directed Search: A Case Study of Job-Shop Scheduling", (PhD Thesis), Technical Report, Robotics Institute, Carnegie-Mellon University, Pittsburgh PA.

Fox, M.S., Allen, B.P., and Strohm, G.A., (1982), "Job-Shop Scheduling: An Investigation in Constraint-Directed Reasoning." *Proceedings of the 2nd National Conference on Artificial Intelligence*,Pittsburgh, Pennsylvania, August 1982.

Fox M.S., B. Allen, S. Smith, and G. Strohm, (1983a), "ISIS: A Constraint-Directed Search Approach to Job-Shop Scheduling", *Proceedings of the IEEE Computer Society Trends and Applications*, National Bureau of Standards, Washington DC. An expanded version appeared as technical report CMU-RI-TR-83-3, Robotics Institute, Carnegie-Mellon University, Pittsburgh PA.

Fox M.S., S. Lowenfeld, and P. Kleinosky, (1983b), "Techniques for Sensor-Based Diagnosis", *Proceedings of the International Joint Conference on Artificial Intelligence*, Karlsruhe, West Germany, August 1983.

Fox, M.S., Allen, B.P., Smith, S.F., and Strohm,G.A.,(1983), "Future Knowledge-based Systems for Factory Scheduling." *Proceedings of CAM-I 12th Annual Meeting and Technical Conference*, Dallas, Texas, 1983.

Fox M., and S. Smith, (1984), "ISIS: A Knowledge-Based System for Factory Scheduling", *International

Journal of Expert Systems, Vol. 1, No. 1, to appear.

Haley P. and J. McDermott, (1984), "PTRANS: A Rule-Based Management Assistant", Technical Report, Computer Science Dept., Carnegie-Mellon University, Pittsburgh PA, in preparation.

Kosy D.and V. S. Dhar, (1983), "Knowledge-Based Support System for Long Range Planning", Technical Report, Robotics Institute, Carnegie-Mellon University, Pittsburgh, Pennsylvania, December, 1983.

Langley, P., G. Bradshaw, and H. A. Simon, (1982) "Data-driven and Expectation-driven Discovery of Empirical Laws", *Proceedings of the Fourth National Conference of the Canadian Society for Computational Studies of Intelligence*, 1982.

Langley, P., (1982), "Strategy Acquisition Governed By Experimentation", *Proceedings of the European Conference on Artificial intelligence*, 1982.

Langley, P., (1982), "A Model of Early Syntactic Development", *Proceedings of the 20th Annual Conference of the Society for Computational Linguistics*, 1982.

Langley, P., (1982), "Language Acquisition Through Error Recovery", Cognition and Brain Theory, 1982.

Langley, P.,(1983), "Learning Search Strategies Through Discrimination", To appear in International Journal of Man-Manchine Studies,1983.

Langley, P. (1983), "A General Theory of Discrimination Learning", To appear in Klahr, D., Langley, P., and Neches, R. (Eds.), Production system models of learning and development, Cambridge, Mass, MIT Press, 1983.

Langley, P., (1983), "Exploring the Space of Cognitive Architectures", To appear in Behavior Research Methods and Instrumentation, 1983.

Langley, P., G. Bradshaw, J. Zytkow, and H. A. Simon, (1983), "Three Facets of Scientific Discovery", IJCAI, 1983.

Langley, P., "Representational Issues in Learning", Computer, special issue on representation.

Morton T., and R. Rachamadugu, (1982), "Myopic Heuristics for the Single Machine Weighted Tardiness Problem", Technical Report, Robotics Institute, Carnegie-Mellon University, Pittsburgh, PA.

Rachamadugu, Ram Mohan V., (1982), "Myopic Heuristics in Job Shop Scheduling" Ph.D. Thesis, Graduate School of Industrial Administration.

Rachamadugu R., A. Vepsalainen, and T. Morton, (1982), "Scheduling in Proportionate Flowshops", Technical Report, Robotics Institute, Carnegie-Mellon University, Pittsburgh, PA.

Reddy Y.V. and M.S. Fox, (1982), "KBS: An Artificial Intelligence Approach to Flexible Simulation", CMU-RI-TR-82-1, Robotics Institute, Carnegie-Mellon University, Pittsburgh PA.

Reddy Y.V., and M.S. Fox, (1983), "INET: A Knowledge-Based Simulation Approach to Distribution Analysis",

Proceedings of the IEEE Computer Society Trends and Applications, National Bureau of Standards, Washington DC.

Sleeman, D., P. Langley, and T. Mitchell, (1982), "Learning from Solution Paths: An approach to the Credit Assignment Problem", AI Magazine, Spring, 1982.

Smith S., (1983), "Exploiting Temporal Knowledge to Organize Constraints", Technical Report, Robotics Institute, Carnegie-Mellon University, Pittsburgh, PA.

Smith S., (1983), Flexible learning of Problem Solvind Heuristics Through Adaptive Search", *Proceedings IJCAI-83*, Karlsruhe, West Germany, August, 1983.

Wright J.M., (1984), "Issues in the Rxepresentation of Knowledge for Simulation and Diagnosis", MS Thesis, Robotics Institute, Carnegie-Mellon University, Pittsburgh PA, in preparation.

Wright J.M., and Fox M.S., (1983), "SRL: Schema Representation Language", Technical Report, Robotics Institute, Carnegie-Mellon University, Pittsburgh PA.

INDEX

D

Geometry, 92, 194
Global area monitors, 180
GO, 144, 163
Goal specifications, 60
Graphic
    aids, 178
    simulators, 98
Graphical mapping, 117
Graphics, 49, 83, 107
    interactive, 107
    screens, 184
Gray levels, 203, 205
Gray scale, 145, 147, 150, 152
    scaling, 145
Group technology, 16, 19, 23-24, 42, 77-78, 87, 94, 123, 139
    codes, 79
    software, 105
Growth potential, 211
GT, See: Group technology
Guided vehicles, 173, 183

H

Hardware, 26, 107, 117, 143, 145, 150-152, 212-213
    computer, 184
    image processing, 200, 203
Heuristics, 37, 50, 61, 73-74, 80, 105, 119-120, 122, 124, 126, 136, 138-140, 194
    knowledge, 113
    searches, 95, 167
Hierarchical scheduling, 122, 240
Histograms, 145, 153-154
History, 135, 216
Human
    abilities, 13
    beings, 9, 211, 216
    expertise, 118, 162
    experts, 139, 228
    input, 234
    inspectors, 194
    intelligence, 37
    operators, 138
    performance, 37, 162
    reasoning, 205
    schedulers, 120, 126, 129
Human-computer interaction, 119
Humanlike intelligence, 102
Hybrid systems, 226

I

Identification, 144
IF-THEN statements, 73, 211, 218-219
Image
    acquisition, 143, 146
    analysis, 143
    digitizers, 199
    generation, 200
    operations, 143-144
    resolution, 144
    understanding, 143-144, 146
Image processing, 143, 146, 151-153, 203
    automated, 205
    hardware, 144, 199-200

subsystems, 202
Imaging devices, 143
Implementation, 75, 95, 173
Implementors, 116
Industrial control, 144, 216-220
Industrial scheduling, 123
Industry, 140, 144
    manufacturing, 11
Inference engines, 37, 43, 177, 211, 218-219, 227-229, 234
Inferencing
    forward, 115
    rules, 211-212
    speed, 231
    systems, 229-230
Inflation, 11
Information, 12
    gathering modules, 78-79
    integration, 11-12, 22
    loop, 76-77
    management systems, 14
    processing, 30
    system architecture, 15
    tracking, 246
Initial graphics exchange standard, 117
Initial graphics exchange specification, 193
Innovations, 27-28
Input, 33, 127, 153, 168
    convertors, 180
    data, 98
    interfaces, 97
Input/output, 43
    facilities, 37
Inspection, 33, 71, 143-145, 147, 196-207, 218
    centers, 182
    costs, 190
    data, 193
    dimensional, 174
    operations, 207
    parts, 144
    processes, 205
    results, 176
    systems, 195, 201
    vision, 189-195
Integration, 47, 71, 170, 185
Integration information, 11-12, 22
Intellectual works, 30
Intelligence, 17
    humanlike, 102
Intelligent
    conveying, 71
    design invention systems, 53
    management systems, 236-250
    vision, 49, 143-148
    work, 30
Interactive
    graphics, 107
    scheduling, 239
    systems, 117
    video disks, 139
Interface, man-machine, 116
Interfaces, 97
Interpretation, 23, 38, 137
Interpreters, 178
Invention, 49-67
Inventive design, 65
Inventory, 11, 144, 161, 176
    management, 121
Investments, 161, 164, 168
Irradiation, 200

Milling, 91
Miniaturization, 146
MIS, See: Management information systems
Model perusal, 239
Models, 12, 29, 33, 38, 50, 73-75, 82, 236, 241, 243
  cognitive, 65
  mathematical, 119
  optimization, 12
  organizational, 237-238
  planning, 12, 73
  prototype, 42, 47
  simulation, 119, 242
Modules, 73, 146, 189, 193-194
  decomposition, 124
  information gathering, 78
  sequences, 78
Monitoring, 38-39, 137, 244-245
Multiaxis robots, 189
Multiexpert systems, 168-170

N

Natural language, 8-9, 12, 22, 38, 66, 105, 110, 211, 216-217, 219, 237, 246
  interface, 212
  processing, 14, 108
  software, 212
NC, See: Numerical control
Net present worth, 161
Networks, 111-112, 114
Nodes, 108, 111
NO GO, 144, 163
Numerical control
  machining, 76
  part programs, 182
  programmers, 77-78

O

Object analysis, 145
Object recognition, 144
Observations, 228
On-line planning, 103
Operation planning, 76-88
Operations, 111
Operation sequencing, 77
Operator guidance systems, 213
Operators, 29, 94-95
Optical character recognition, 145
Optimization, 12, 237
Order definition, 239
Organizational models, 237-238
Orientation, 151
Output, 32-33, 127, 147, 180, 194
  interfaces, 97-98
Overhead costs, 76
Overlap flagging, 240

P

Part
  analysis, 143
  descriptions, 73
  features, 78

geometry, 73, 79
inspection, 144
loading, 177
positioning, 218
programmers, 90
programming, 89-101
Pattern matching, 167
Pattern recognition, 94, 144
Payback, 139
Perception, 218
Personal computers, 143, 162, 231
Perspective transformation, 156
Physical knowledge, 55, 65
Physics, 54
Pick-and-place robots, 145
Pixels, 143, 148, 152-154, 203
Planners, 79, 115, 139
Planning, 23, 38-39, 41, 137
  capacity, 121
  financial, 242
  machining, 77
  manufacturing, 11, 115
  master, 121
  models, 12, 73
  on-line, 103
  operation, 77
  process, 25, 31-33, 71-72, 76, 102
  production, 19, 71-75, 121, 233
  systems, 87, 102-118
Plant
  monitoring, 241
  organization, 239
Power system restoration, 215
Prediction, 137
Primitives, conceptual, 238
Printed circuit boards, 104, 115, 145
Problem solving, 56, 59, 73, 76-77, 91, 218
  algorithms, 167
Problems, 135
Procedural-language programming, 79
Process
  alarms, 213
  control, 7, 211-220, 232, 236
  diagnosis, 246
  manufacturing, 46
  optimization, 219
  planners, 71-72, 78
  planning, 24-25, 31-33, 41, 71-72, 74, 76-77, 81-82, 87, 173, 241
  plans, 46, 102, 104
  technologies, 28
Producibility analysis system, 105
Product
  costs, 190
  descriptions, 107
  design, 39-40, 42
  management, 244
  quality, 103
Production, 29, 72, 219, 231, 236
  control, 118
  engineering, 73
  lines, 146
  management, 238
  managers, 233
  rules, 73, 93, 111, 229, 232
  scheduling, 74, 119
Production planning, 40-41, 71-75, 121-122, 124-125, 127, 226-227, 233-234
Productivity, 13-15, 27, 76, 144, 220
Profitability, 161-162

Surveillance, 144
Symbols, 140, 216
Synchronization, 33
System
    components, 78
    configurations, 150
    designers, 121
    implementors, 116
    managers, 173-174, 184
    shells, 228
Systems, 28
    control, 238
    expert, 32, 73, 113, 226-235
    houses, 146
    inferencing, 230
    integration, 173
    simulation, 237
    support, 242
    vision, 195

T

Teach modes, 207
Technology, 31-32, 140
    innovations, 26
    manufacturing, 25-34
    parameters, 81
    transfer, 13
Testing, 198
Three-dimensional
    milling, 95
    networks, 104
    parts, 145
    research, 147
Throughput, 117, 139, 144, 161-162
Time-of-flight techniques, 147
Tokens, 106
Tool
    inventory, 82
    paths, 100
    planning, 24
    selection, 78
Training tools, 224
Translation
    data, 184
    module, 189
    programs, 191, 193
Transportability, 234
Troubleshooters, 225
Turnkey systems, 144, 146
Two-dimensional, 143
    objects, 147
    problems, 91-92, 95
    strategies, 95
Two-way communication, 173

U

Users, 212
    interface, 218

V

Variant process planning, 72
Vendors, 231
Video
    image analysis, 196, 198
    protocol, 117
    recorders, 199
Vision, 7, 13
    industry, 146
    inspection, 189-195
    intelligence, 146
    modules, 144
    robot, 149-158
    systems, 143, 149, 151-152, 156-158, 191-192, 194-195, 216-218
    technology, 144
Visual
    imagination, 65
    recognition, 212
Voice recognition, 9, 212

W

Welding, 218
Windowing, 144
Work
    area definition, 239
    cells, 149-150
    materials, 78
Work-in-process, 11
Work-in-progress, 139
Workplace layout, 139
Works, intellectual, 30
Workspace, 218-219
Workstations, 104, 117, 183-184

X

X-rays, 200

Z

Zooming, 205

DATE DUE

SE